REVIEW OF THE

Research Program of the PARTNERSHIP for a NEW GENERATION of VEHICLES

Seventh Report

Standing Committee to Review the Research Program of the Partnership for a New Generation of Vehicles

Board on Energy and Environmental Systems
Division on Engineering and Physical Sciences
Transportation Research Board
National Research Council

NATIONAL ACADEMY PRESS
Washington, D.C.

National Academy Press • 2101 Constitution Avenue, N.W. • Washington, DC 20418

NOTICE: The project that is the subject of this report was approved by the Governing Board of the National Research Council, whose members are drawn from the councils of the National Academy of Sciences, the National Academy of Engineering, and the Institute of Medicine. The members of the committee responsible for the report were chosen for their special competences and with regard for appropriate balance.

This report and the study on which it is based were supported by Contract No. DTNH22-00-G-07519. Any opinions, findings, conclusions, or recommendations expressed in this publication are those of the author(s) and do not necessarily reflect the views of the organizations or agencies that provided support for the project.

Library of Congress Control Number: 2001094462
International Standard Book Number 0-309-07603-X

Available in limited supply from:
Board on Energy and Environmental Systems
National Research Council
2101 Constitution Avenue, N.W.
HA-270
Washington, DC 20418
202-334-3344

Additional copies are available for sale from:
National Academy Press
2101 Constitution Avenue, N.W.
Box 285
Washington, DC 20055
800-624-6242 or 202-334-3313 (in the Washington metropolitan area)
http://www.nap.edu

Printed in the United States of America

THE NATIONAL ACADEMIES

National Academy of Sciences
National Academy of Engineering
Institute of Medicine
National Research Council

The **National Academy of Sciences** is a private, nonprofit, self-perpetuating society of distinguished scholars engaged in scientific and engineering research, dedicated to the furtherance of science and technology and to their use for the general welfare. Upon the authority of the charter granted to it by the Congress in 1863, the Academy has a mandate that requires it to advise the federal government on scientific and technical matters. Dr. Bruce M. Alberts is president of the National Academy of Sciences.

The **National Academy of Engineering** was established in 1964, under the charter of the National Academy of Sciences, as a parallel organization of outstanding engineers. It is autonomous in its administration and in the selection of its members, sharing with the National Academy of Sciences the responsibility for advising the federal government. The National Academy of Engineering also sponsors engineering programs aimed at meeting national needs, encourages education and research, and recognizes the superior achievements of engineers. Dr. Wm. A. Wulf is president of the National Academy of Engineering.

The **Institute of Medicine** was established in 1970 by the National Academy of Sciences to secure the services of eminent members of appropriate professions in the examination of policy matters pertaining to the health of the public. The Institute acts under the responsibility given to the National Academy of Sciences by its congressional charter to be an adviser to the federal government and, upon its own initiative, to identify issues of medical care, research, and education. Dr. Kenneth I. Shine is president of the Institute of Medicine.

The **National Research Council** was organized by the National Academy of Sciences in 1916 to associate the broad community of science and technology with the Academy's purposes of furthering knowledge and advising the federal government. Functioning in accordance with general policies determined by the Academy, the Council has become the principal operating agency of both the National Academy of Sciences and the National Academy of Engineering in providing services to the government, the public, and the scientific and engineering communities. The Council is administered jointly by both Academies and the Institute of Medicine. Dr. Bruce M. Alberts and Dr. Wm. A. Wulf are chairman and vice chairman, respectively, of the National Research Council.

STANDING COMMITTEE TO REVIEW THE RESEARCH PROGRAM OF THE PARTNERSHIP FOR A NEW GENERATION OF VEHICLES

CRAIG MARKS (Chair), NAE,[1] AlliedSignal (retired), Bloomfield Hills, Michigan
VERNON P. ROAN (Vice Chair), University of Florida, Gainesville
WILLIAM AGNEW, NAE, General Motors Research Laboratories (retired), Washington, Michigan
KENNERLY H. DIGGES, George Washington University, Washington, D.C.
W. ROBERT EPPERLY, Epperly Associates, Mountain View, California
DAVID E. FOSTER, University of Wisconsin, Madison
NORMAN A. GJOSTEIN, NAE, University of Michigan, Dearborn
DAVID F. HAGEN, Ford Motor Company (retired), Dearborn, Michigan
JOHN B. HEYWOOD, NAE, Massachusetts Institute of Technology, Cambridge
FRITZ KALHAMMER, Electric Power Research Institute (retired), Palo Alto, California
JOHN G. KASSAKIAN, NAE, Massachusetts Institute of Technology, Cambridge
HAROLD H. KUNG, Northwestern University, Evanston, Illinois
DAVID F. MERRION, Detroit Diesel Corporation (retired), Brighton, Michigan
JOHN SCOTT NEWMAN, NAE, University of California, Berkeley
ROBERTA NICHOLS, NAE, Ford Motor Company (retired), Plymouth, Michigan
F. STAN SETTLES, NAE, University of Southern California, Los Angeles

Committee Subgroup on Systems Analysis and Electrical and Electronic Systems

JOHN G. KASSAKIAN (Chair)
JOHN B. HEYWOOD
JOHN SCOTT NEWMAN
ROBERTA NICHOLS
F. STAN SETTLES

Committee Subgroup on Batteries

FRITZ KALHAMMER (Chair)
HAROLD H. KUNG
JOHN SCOTT NEWMAN
VERNON P. ROAN

[1]NAE = National Academy of Engineering

Committee Subgroup on Fuels

W. ROBERT EPPERLY (Chair)
DAVID E. FOSTER
DAVID F. MERRION
ROBERTA NICHOLS

Committee Subgroup on Fuel Cells

VERNON P. ROAN (Chair)
FRITZ KALHAMMER
HAROLD H. KUNG
JOHN SCOTT NEWMAN

Committee Subgroup on Internal Combustion Engines and Emissions Control

DAVID E. FOSTER (Chair)
WILLIAM AGNEW
W. ROBERT EPPERLY
DAVID F. HAGEN
JOHN B. HEYWOOD
DAVID F. MERRION

Committee Subgroup on Materials and Safety

NORMAN A. GJOSTEIN (Chair)
KENNERLY H. DIGGES
DAVID F. HAGEN
F. STAN SETTLES

Committee Subgroup on Cost Analysis

CRAIG MARKS (Chair)
WILLIAM AGNEW
DAVID F. HAGEN
DAVID F. MERRION
F. STAN SETTLES

Project Staff

JAMES ZUCCHETTO, Responsible Staff Officer, Board on Energy and Environmental Systems (BEES)
NAN HUMPHREY, Senior Program Officer, Transportation Research Board
SUSANNA E. CLARENDON, Senior Project Assistant and Financial Associate (BEES)

BOARD ON ENERGY AND ENVIRONMENTAL SYSTEMS

ROBERT L. HIRSCH (Chair), RAND, Arlington, Virginia
RICHARD E. BALZHISER, NAE,[1] Electric Power Research Institute, Inc. (retired), Menlo Park, California
DAVID L. BODDE, University of Missouri, Kansas City
PHILIP R. CLARK, NAE, GPU Nuclear Corporation (retired), Boonton, New Jersey
WILLIAM L. FISHER, NAE, University of Texas, Austin
CHRISTOPHER FLAVIN, Worldwatch Institute, Washington, D.C.
HAROLD FORSEN, NAE, Foreign Secretary, National Academy of Engineering, Washington, D.C.
WILLIAM FULKERSON, Oak Ridge National Laboratory (retired) and University of Tennessee, Knoxville
MARTHA A. KREBS, California Nanosystems Institute, Los Angeles, California
GERALD L. KULCINSKI, NAE, University of Wisconsin, Madison
EDWARD S. RUBIN, Carnegie Mellon University, Pittsburgh, Pennsylvania
ROBERT W. SHAW JR., Aretê Corporation, Center Harbor, New Hampshire
JACK SIEGEL, Energy Resources International, Inc., Washington, D.C.
ROBERT SOCOLOW, Princeton University, Princeton, New Jersey
KATHLEEN C. TAYLOR, NAE, General Motors Corporation, Warren, Michigan
JACK WHITE, Association of State Energy Research and Technology Transfer Institutions, Falls Church, Virginia
JOHN J. WISE, NAE, Mobil Research and Development Corporation (retired), Princeton, New Jersey

Staff

JAMES ZUCCHETTO, Director
RICHARD CAMPBELL, Program Officer
ALAN CRANE, Program Officer
MARTIN OFFUTT, Program Officer
SUSANNA CLARENDON, Financial Associate
PANOLA GOLSON, Project Assistant
ANA-MARIA IGNAT, Project Assistant
SHANNA LIBERMAN, Project Assistant

[1] NAE = National Academy of Engineering

Acknowledgments

The committee wishes to thank all of the members of the Partnership for a New Generation of Vehicles, who contributed a significant amount of their time and effort to this National Research Council (NRC) study by giving presentations at meetings, responding to requests for information, or hosting site visits. The committee also acknowledges the valuable contributions of other organizations that provided information on advanced vehicle technologies and development initiatives. Finally, the chair wishes to recognize the committee members and the staff of the NRC Board on Energy and Environmental Systems for their hard work in organizing and planning committee meetings and their individual efforts in gathering information and writing sections of the report.

This report has been reviewed by individuals chosen for their diverse perspectives and technical expertise, in accordance with procedures approved by the NRC's Report Review Committee. The purpose of this independent review is to provide candid and critical comments that will assist the authors and the NRC in making the published report as sound as possible and to ensure that the report meets institutional standards for objectivity, evidence, and responsiveness to the study charge. The content of the review comments and draft manuscript remain confidential to protect the integrity of the deliberative process. We wish to thank the following individuals for their participation in the review of this report: Charles Amann (NAE), Gary L. Borman (NAE), Pat Flynn (NAE), Robert A. Frosch (NAE), Harvard University, Roger McClellan (IOM), Jerome G. Rivard (NAE), Global Technology and Business Development, Dale F. Stein (NAE), R. Rhoads Stephenson, and Supramaniam Srinivasan, Princeton University.

Although the reviewers listed above have provided many constructive comments and suggestions, they were not asked to endorse the conclusions and recommendations, nor did they see the final draft of the report before its release. The review of this report was overseen by Trevor Jones, Biomec, Inc., appointed by the NRC's Division on Engineering and Physical Sciences, and Gary Byrd, consulting engineer, appointed by the Report Review Committee, who were responsible for making certain that an independent examination of the report was carried out in accordance with institutional procedures and that all review comments were carefully considered. Responsibility for the final content of this report rests entirely with the authoring committee and the institution.

Contents

Tables and Figures

TABLES

FIGURES

REVIEW OF THE

Research Program of the PARTNERSHIP for a NEW GENERATION of VEHICLES

Seventh Report

Executive Summary

This is the seventh report by the National Research Council Standing Committee to Review the Research Program of the Partnership for a New Generation of Vehicles (PNGV). The PNGV program is a cooperative research and development (R&D) program between the federal government and the United States Council for Automotive Research (USCAR), whose members are DaimlerChrysler Corporation, Ford Motor Company, and General Motors Corporation (GM). The program addresses improvements in national competitiveness in manufacturing and in the implementation of energy-saving innovations in passenger vehicles. In addition, it seeks to develop a new generation of vehicles by setting a stretch goal to achieve fuel economy up to three times (80 miles per gallon [mpg] gasoline equivalent) that of comparable 1994 family sedans without sacrificing size or utility or increasing the cost of ownership. The purpose of this program is to conceive, develop, and implement new technologies capable of significantly reducing the petroleum consumption and carbon dioxide emissions of the U.S. automobile fleet. The founders recognized that, to have substantial impact, this new generation of vehicles must be sold in high volume. This, in turn, requires that the vehicles meet or exceed all emission and safety requirements and offer all of the characteristics that result in strong customer appeal.

This report contains the committee's assessment of the overall balance and adequacy of the PNGV research program to meet its technical goals and the program's efforts to develop commercially feasible low-emission propulsion systems. The committee also comments on significant changes that have occurred since the inception of the PNGV program and how these changes might influence this program.

PROGRESS AND MAJOR ACHIEVEMENTS

The PNGV program has overcome many challenges and has forged a useful and productive partnership of industry and government participants. In addition to the cooperative program, substantial proprietary industry R&D activity has been generated. Teams of industry and government representatives have addressed formidable technical issues and made significant progress on many of them despite the complexity of managing an inter-disciplinary program involving three competing companies, several government agencies, and significant government budget constraints. The program concept cars introduced in January and February of 2000 are important evidence of these activities, but the ongoing R&D program, much of which is summarized in the following sections, is equally significant.

The following summarizes activities for meeting goals 1, 2, and 3 of the program.

Goal 1

The manufacturing competitiveness goal, Goal 1, addresses the need to develop improved manufacturing processes for conventional vehicles, as well as the new-generation vehicles and their components. A wide array of manufacturing issues has been addressed in the cooperative program. Projects to reduce the cost and improve the quality of aluminum structures, drill holes more rapidly, and improve leak testing were completed in 2000-2001. Several projects to facilitate the manufacture of lower-cost, lighter-weight vehicle bodies have been proposed for funding in fiscal year 2002. Manufacturing considerations are being addressed for many of the new components that will be required by the radically different hybrid-electric-vehicle power trains being developed. Also, several longer-term and higher-risk manufacturing projects are at the proposal stage.

Since a large proportion of the components needed to assemble automobiles comes from suppliers, the need for manufacturing improvements extends well beyond the automobile manufacturers themselves. Suppliers are already involved in some PNGV activities, but the PNGV manufacturing program would benefit from expansion of these supplier activities.

Goal 2

The purpose of Goal 2 is to speed the introduction of new technology generated by PNGV R&D into production vehicles. Several manufacturing and engineering analysis tools developed by the program are in use, and significant applications of lightweight materials have been introduced in production vehicles.

The most striking Goal 2 achievement is the announced plans by all three automobile companies to introduce hybrid power trains during the next two to three years in both pickup trucks and sport utility vehicles in a variety of configurations. The reduction in fuel consumption will range from 10 to 30 percent, twice the amount that would be saved if the same percentage reduction were obtained by applying hybrid technology to a mid-size car that initially had two times the fuel economy (mpg) of these trucks. The committee commends the automobile companies for this commitment to produce vehicles that will significantly reduce the total fuel consumption of the light-duty vehicle fleet even with an increase in sales.

Goal 3

Goal 3 has provided an extremely challenging focus for the program: to develop within 10 years (by 2004) vehicles that will achieve up to three times the fuel efficiency of comparable 1994 family sedans while retaining the features that make them marketable and affordable. The year 2000 concept-vehicle milestone was met when the three manufacturers each introduced concept cars: the DaimlerChrysler ESX3, the Ford Prodigy, and the General Motors Precept, as detailed in the last committee report. All three concept vehicles incorporate hybrid-electric power trains designed around small, turbocharged, compression-ignition direct-injection (CIDI) engines, using diesel fuel, which shut down when the vehicles come to rest. All employed the significant technical advances developed in the PNGV program to reduce the energy requirements for propelling the vehicle (e.g., reduced mass, aerodynamic drag) and for supplying auxiliary loads (e.g., heating, air conditioning). Each company took a different approach to the design of these cars, which resulted in different remaining challenges to meet the fuel economy and affordability targets, but all of the cars operate on diesel fuel. These cars provide a valuable measure of how challenging it will be to meet all the components of Goal 3 simultaneously.

The next major Goal 3 milestone of the PNGV program as currently structured is the development of production-prototype cars by 2004. Each car company is in the planning stage for this activity, and the approach that each may take is not clear. Validation of production readiness for a new car requires immense resources compared to the preceding R&D activities. For these resources to be justified, the car must be one that is included in the production plans of each manufacturer, plans that are, of course, proprietary. In order for the committee to evaluate the PNGV program in context, each year the car companies have shared proprietary information with the committee. As work progresses toward production prototypes, more of it becomes proprietary and this limits the detail about Goal 3 activity that can be reported by the committee in this and future reports.

Vehicle Engineering, Structural Materials, and Safety

The PNGV concept vehicles made public last year all made extensive use of lightweight materials and new body construction techniques to achieve major reductions (20 to 31 percent) in curb weight. The high cost of these lightweight materials and the associated manufacturing costs represent a significant part of the affordability challenge faced by the program. More than 30 materials projects have been established to attack the technical challenges identified. In addition, the car companies each have proprietary programs, and the American Iron and Steel Institute has embarked on a second-phase advanced vehicle concept car aimed at identifying affordable ways to reduce weight.

As progress is made on these projects the benefits of lighter-weight construction will be achieved in production vehicles. The PNGV program has developed lower-cost, lightweight-material production processes such as continuous casting of aluminum sheet, powder-metal processes for aluminum-metal matrix composites, and a microwave process for producing carbon fiber. Vehicle production programs using these materials probably will be necessary to provide material suppliers with the incentive to invest in these new processes.

The newly formed PNGV Safety Working Group is addressing safety issues that have been raised by the concept-car designs. The crashworthiness of lighter-weight vehicles in car-to-car accidents is an issue being studied. While the stated goal is to meet present and future Federal Motor Vehicle Safety Standards, it is recognized that these are minimum standards. The purpose of the Safety Working Group is to identify and sponsor research directed at the unique safety characteristics of PNGV vehicles in order to help ensure the marketability of vehicles employing these new technologies.

Four-Stroke Direct-Injection Engines and Fuels

The CIDI engine operating on diesel fuel, chosen for its high efficiency, continues to be the major focus of PNGV power-plant development for near-term application. Current PNGV activity centers on the challenge of meeting new emission standards and is being pursued in engine combustion, exhaust-gas after-treatment, and fuels development programs. Aggressive emission reduction targets have been set for the program to meet through the year 2007. As noted in last year's report (NRC 2000), these emission targets, driven by the newly promulgated Tier 2 emission standards, are now much more stringent than they were at the outset of the PNGV program.

In the combustion program, diagnostic techniques for measuring cylinder-to-cylinder distribution of recirculated exhaust gas (a key NO_x control measure) and for in-cylinder measurement of particulate particle size and number have made progress this year. Advanced simulation techniques have also been developed and validated, with the promise of these techniques becoming useful as an optimizing tool for engine design.

NO_x exhaust-gas after-treatment is being pursued using lean-NO_x absorber catalysts, selective catalytic reduction systems using urea, and nonthermal plasma catalytic systems. Development is in an early stage, and all systems result in a fuel economy penalty, some estimated as low as 0.5 percent and others as high as 8 percent. Nitrogen oxide traps also are being tested with the current conclusion that, because of extreme sensitivity to sulfur poisoning, they may require simultaneous use of a sulfur trap. Particulate reduction will require yet another trap and an effective regeneration mechanism.

The engine-fuel interactions program was focused on the effects of fuel chemistry and physical properties on engine performance and emissions. Results to date indicate that the fuel does have an effect on engine-out particulate and NO_x emissions, but that these effects are not large enough to eliminate the need for substantial after-treatment. The sulfur level in fuels will have a significant effect on both engine-out particulates and the performance of after-treatment systems. Quantification of these relationships remains a priority.

Fuel Cells

Fuel cells continue to show promise of high efficiency and very low emissions with continuous progress toward targets that are very difficult to meet for any general-purpose, high-volume automotive application. There are many substantial barriers remaining to be overcome prior to the realization of a mass-manufactured consumer vehicle. These barriers include performance as well as physical, fuel-related, and cost issues. In the short term it appears that some limited-production fleet vehicles will operate on pure hydrogen stored onboard the vehicle, which results in a simpler and less expensive system for the vehicle; however, for the foreseeable future, high-volume, general-purpose vehicles likely will require the fuel cell system to be combined with an onboard reformer that produces hydrogen from a liquid fuel. The efficiency of these liquid-fuel reformers is a critical issue: Current prototype reformers significantly degrade the overall fuel cell system efficiency.

This year a major program milestone was the demonstration of two integrated gasoline-fueled 50-kW fuel cell systems. The projected size, weight, and cost of these systems are short of the original year 2000 targets by a large margin, but these systems represent encouraging progress and will help define the goals for component development that will improve system performance. Significant improvements have been demonstrated in many of the components: fuel processors, heat exchangers, catalysts, bipolar plates, and complete stacks. In addition, large proprietary programs are under way both by the automobile companies and potential fuel cell suppliers, and these programs are driving component improvements in all of these areas.

Batteries

Research and development on batteries continues to focus on nickel metal hydride, lithium-ion, and more recently, lithium-polymer designs. Full-scale systems employing each of these designs have demonstrated their capability to meet the key technical performance targets for hybrid-vehicle applications during the past year, but the battery life challenge has become even more severe. The technical team has adopted targets that correspond to a 15-year battery calendar and operating life. The battery cost targets seem to be very aggressive. The costs of batteries in the currently marketed Japanese hybrid vehicles exceed these PNGV targets by a factor of five or more. These targets should be re-examined in the context of the prospects for meeting cost targets for the other key hybrid-vehicle subsystems.

Supporting basic research has helped define fundamental failure mechanisms of lithium-ion cells and the cause of thermal runaway. New cells have shown life improvement in elevated-temperature accelerated tests, and more realistic calendar life testing methods are being developed.

Power Electronics and Electrical Systems

Both the power electronics and electric motor programs are focused on reducing the cost of these components. Three contractors for the power electronics module have each executed a detailed economic gap analysis with their suppliers to identify ways to ensure that their cost target can be met. These analyses have provided detailed plans for material, labor, and overhead cost reductions, and these plans provide reasonable confidence that the goals can be met. One of the electric motor contractors is pursuing an axial gap permanent-magnet motor design, and the other is using a more conventional radial gap induction machine.

Research programs at the national laboratories and at universities continue to develop promising technologies for essential electronic and motor materials and components. These include silicon-carbide-power semiconductors, carbon-foam thermal materials, high-energy magnets, and low-cost, high-dielectric-constant materials for capacitors.

MAJOR BARRIERS

As noted, significant progress continues to be made by the research being performed in the PNGV partnership and in the many proprietary programs being carried out by the individual partners in USCAR. Nevertheless, the committee believes it is unlikely that all of the elements of Goal 3, including three-times fuel economy, will be met in production-prototype vehicles in 2004. While the bulk of the requirements (e.g., performance, comfort, cargo space, utility, and safety) can be met, the combination of 80 mpg and affordability appears out of reach. In addition, the recently promulgated Environmental Protection Agency (EPA) Tier 2 emission requirements will require radically better emission control technology. It also appears that the required after-treatment devices may significantly degrade the efficiency of the CIDI engine and increase its cost. Fuel issues also pose significant questions that affect the viability of widespread automobile use of the CIDI engine and, longer term, of fuel cells, which require a supply of hydrogen or a fuel that can be converted to hydrogen.

Cost Challenge

High prospective cost is a serious problem in almost every area of the PNGV program. Lightweight body construction, CIDI engines, batteries, and electronic control systems all represent increases in vehicle cost. Needed emission exhaust-gas after-treatment devices are not well defined at this point, but they will most certainly be more expensive than systems currently employed. The major effort to date has been to achieve the technical targets for these components, and the concept cars demonstrate the significant progress made; however, none of these cars in their present forms represents an affordable set of components compatible with similar mission vehicles.

Cost targets have always been in place for the major components, but it has not been clear to the committee that even if these targets were achieved an affordable vehicle would result. This year a new cost-modeling effort has been started to address this vital subject. The plan is to develop a tool that will help the technical teams direct their pre-competitive R&D efforts and help suppliers find ways to reduce the gap between current costs and those needed to get to production feasibility. The committee compliments the PNGV for getting this effort under way.

As noted earlier, affordability is the linchpin of the PNGV program. For the benefits PNGV intended to be realized, the economics must favor large-scale purchases of these vehicles.

Exhaust Emissions Trade-off

The last committee report (NRC, 2000) noted that the Tier 2 NO_x and particulate matter (PM) emission standards could preclude the early introduction and widespread use in the United States of CIDI engines for passenger cars. Without the CIDI engine the fuel economy of near-term PNGV cars could drop by as much as 25 percent, the approximate difference in fuel economy between a CIDI and a homogeneous-charge, spark-ignition engine. Although, as detailed later in this report, significant progress is being made in developing exhaust after-treatment systems for CIDI engines, these devices make this power plant less attractive by increasing its fuel consumption and cost. Alternative power plants that can meet the Tier 2 emission standards will, in all likelihood, have substantially higher fuel consumption and carbon dioxide emissions. This raises the obvious policy question of the relative importance to the nation of decreasing fuel consumption and carbon dioxide emissions compared with the need to tighten the NO_x and PM standards at this time. This trade-off was noted in the last committee report, but the committee is unaware of any subsequent substantive discussion of the issue. Its resolution has obvious implications for the PNGV production-prototype planning process that is now under way.

Fuel Issues

Historically, major improvements in automobile power-plant efficiency and exhaust emissions have required changes in the fuels they use. Notable examples are the high-octane fuel that was required by high-compression-ratio engines and the unleaded fuel required by catalytic converters. Both the CIDI engine and fuel cells being considered by the PNGV are no exception. Successful introduction of either new power plant will be critically dependent on widespread availability of suitable fuels.

The large capital expenditures and long lead time required to manufacture and distribute a significantly modified fuel means that the petroleum industry must be fully aware of the needs well in advance of the production of the first automobile that requires such a fuel. Furthermore, the change must make economic sense for the petroleum companies or be mandated by regulation. In early 2001, the EPA published a regulation requiring refiners to produce highway diesel fuel with a maximum sulfur content of 15 ppm by June 1, 2006 (Federal Register, 2001). This regulation gives the PNGV CIDI development program the challenge of finding an exhaust after-treatment system that will perform and endure with such a fuel, since it is unlikely that fuel with any lower sulfur level will be available in this time frame.

Automotive fuel cell power plants present a much more complicated problem because of the early development stage of these systems. The most efficient and lowest-emission system involves direct hydrogen storage on the vehicle, which requires major infrastructure changes by the energy industry. With a reformer onboard the car, a liquid fuel can be used, and it is hoped that one similar to gasoline will be satisfactory. In the long term, reformers probably will require a fuel tailored for this application to achieve optimum efficiency and minimum emissions.

From this discussion it is clear that a strong, objective, cooperative program between the PNGV participants and the petroleum industry is needed to ensure that the lack of appropriate fuels does not become a major barrier to realizing the goals of the program. It appears that additional priority will be required to advance this goal, as there has been little apparent progress in this area since the committee made a similar recommendation last year.

Fuel Cells

From the inception of PNGV, practical automotive fuel cell power plants have been considered to be well beyond the 2004 time limit of the program.

Nevertheless, because of their potential for high energy efficiency and no onboard emissions of any regulated pollutants when using hydrogen as a fuel, the development of these systems has remained a major part of PNGV. As noted above, progress has been steady, and some important milestones have been met. Nevertheless, the original targets for 2000 for the fuel cell system were not met. At present, it appears that the dates for meeting these targets should be extended substantially. Size and weight need to be reduced by at least a factor of two to meet the 2004 targets, and cost is roughly six times above the target value for a 2004 PNGV-type vehicle. Even with these formidable challenges, based on projections from the major auto manufacturers it appears that some limited-application fuel-cell-powered vehicles may be produced in the 2003-2005 time frame. Even ignoring cost, these vehicles will likely not be suitable for sale to the general public. It is expected that they will operate with onboard hydrogen storage systems and therefore be restricted to fleet use, where limited range and complex refueling issues can be managed.

Large investments are being made in the commercial development of fuel cell power plants for stationary and nonpropulsion mobile applications. These applications are likely to become successful well before the more stringent cost, size, and weight requirements for an automobile power plant can be met. Some of the extensive R&D being performed for these commercial applications and manufacturing experience with them may help the development of a practical automotive system, but the R&D needed to address the requirements of a vehicle power plant is unique. The PNGV program and extensive proprietary work in the car companies are meeting this need.

ADEQUACY AND BALANCE OF THE PNGV PROGRAM

The adequacy and balance of the PNGV R&D program are difficult to assess. Goals 1 and 2 are stated in qualitative terms, and, as noted previously, 80-mpg production prototypes meeting Goal 3 requirements are not likely to be realized in 2004. The last committee report (NRC, 2000) contained an extensive discussion of this subject, including at least three definitions of success that the existing goals allow:

1. The attainment of all aspects of Goal 3;
2. The development of 2004 production-ready vehicles with fuel economy and cost that maximize potential market penetration; and
3. The accelerated application of PNGV-developed technology to production vehicles and the development of much more fuel efficient technology for application beyond 2004.

The committee believes that no reasonable amount of funding would ensure achievement of all aspects of Goal 3, including 80 mpg, the first definition of success, and that has been clear for some time. Breakthrough ideas and talented people are more stringent constraints than money to achieving this goal. Current activities appear to be directed toward the latter two definitions of success; however, no clearly stated objectives have been enunciated by the PNGV. This deficiency needs to be corrected before a meaningful external assessment of the adequacy and balance of the program can be made.

Government funding for the program comes primarily from the U.S. Department of Energy (DOE) advanced automotive technology budget. For fiscal year 2002 this amount was initially proposed to be about $147 million, an increase of about 10 percent from the previous year. At the time of this report the budget process was still under way, but a substantial cut to $100 million has been proposed in the President's budget for the DOE PNGV funding. Other funds identified as supporting the PNGV total about $87 million: in the budgets of the EPA ($27 million), the Department of Commerce ($15 million), and the National Science Foundation ($47 million). Of these latter amounts about three-quarters are only indirectly associated with the program, not directly coordinated with the efforts of the technical teams. The balance of the programs directly coordinated by the technical teams appears to be appropriately weighted toward solving long-range research problems.

Industry funding for "PNGV-related" research has been previously reported to be over $980 million per year for four years of the program, far higher than the 50/50 government-industry matching common in many cooperative programs. A major portion of this funding is in proprietary product programs, the details of which are unavailable to the committee. Furthermore, as the program moves more toward the application of technology to production vehicles, determining the appropriate portion of overall company R&D expenditures that should be associated with the PNGV becomes highly subjective.

THE FUTURE OF THE PNGV

The committee believes that the PNGV program has established a unique and a valuable framework for directing closely coordinated industry and government research efforts toward the development of technologies capable of solving important societal problems. These efforts have resulted in a number of significant technical successes to date. It appears, however, that the current context of the partnership is sufficiently different from that in 1993 to warrant a reconsideration of its specific goals.

The issues addressed by the program are still relevant. The need to reduce the fuel consumption and carbon dioxide emissions of the U.S. automotive fleet is more urgent than ever. Since 1993 there has been a 20 percent increase in the petroleum used in highway transportation, the percentage of U.S. petroleum use derived from imports has risen to 52 percent, and in many parts of the world concerns about the potential for climate change associated with greenhouse gases are even more acute.

On the other hand, during this time, the demand for sport utility vehicles, vans, and pickup trucks in the United States has drastically increased to the point that they now make up 46 percent of new light-duty vehicle sales. This has increased the importance of reducing the fuel consumption of these vehicles compared to that of the typical family sedan. The EPA Tier 2 emission standards in all likelihood will increase the fuel consumption of all new cars and threaten to preclude the widespread introduction of the more efficient diesel engine in light-duty vehicles. Lastly, the changed global structure of the industry has made it much more difficult to make sense of the U.S. competitiveness statement in Goal 1.

In view of these facts and as a new energy policy is being developed for the nation, it is the committee's belief that the priorities and specific goals of the PNGV program should be reexamined. There is a need to update the program goals and technical targets in the context of current and prospective markets. The program would also benefit from a better mechanism that will provide more of a systems approach to identifying issues, planning for issue resolution, and tracking process. The PNGV governance structure contains an Operational Steering Group with high-level representation from both the industry and each of the participating agencies. This provides an opportunity to develop policy trade-offs to ensure that the best interests of the nation are served in economics, the environment, national security, and mobility—an opportunity yet to be realized in any significant way.

SELECTED RECOMMENDATIONS

Recommendation. Taking into consideration the successes, degree of progress, and lessons learned in the PNGV program to date, government and industry participants should refine the PNGV charter and goals to better reflect current societal needs and the ability of a cooperative, precompetitive R&D program to address these needs successfully.

Recommendation. The PNGV should continue the aggressive pursuit and development of lean-combustion exhaust-gas after-treatment systems. The PNGV should also work to develop a detailed systems-modeling effort to quantify the fuel economy penalty associated with using different technologies to meet the emission standards. These efforts should include quantification of the extent to which vehicle hybridization can be used to reduce emissions and the fuel consumption impact of changing the vehicle's primary energy converter.

Recommendation. Because of the potential for near-zero tailpipe emissions and high energy efficiency of the fuel cell, the PNGV should continue research and development efforts on fuel cells even though achievement of performance and cost targets will have to be extended substantially beyond original expectations.

Recommendation. Because affordability is a key requirement of the 2004 production-prototype vehicles, the committee believes that more attention should be paid to the design and manufacturing techniques being worked on by the American Iron and Steel Institute in the Ultralight Steel Auto Body Advanced Vehicle Concept project. These techniques should be applied to aluminum-intensive vehicles, as well as hybrid-material body construction. More broadly, the committee urges a systematic, critical examination of the prospects to achieve cost goals for all key vehicle subsystems and components.

Recommendation. High priority should be given to determining what fuel sulfur level will permit the preferred compression-ignition direct-injection (CIDI) engine and its after-treatment system to meet all regulatory and warranty requirements. An enhanced cooperative effort between the auto and petroleum industries should be undertaken to ensure that the fuels needed commercially will be available on a timely basis.

1

Introduction

On September 29, 1993, President Clinton initiated the Partnership for a New Generation of Vehicles (PNGV) program, a cooperative research and development (R&D) program between the federal government and the United States Council for Automotive Research (USCAR), whose members are DaimlerChrysler Corporation, Ford Motor Company, and General Motors Corporation (GM).[1] This PNGV agreement created a unique industry and government partnership with the broad objective of "strengthening U.S. competitiveness by developing technologies for a new generation of vehicles" (PNGV, 1995). These vehicles were to be conceived using new technologies to produce much lower national gasoline consumption and carbon dioxide emissions. The PNGV Program Plan recognized that, to have substantial impact, these vehicles must be sold in high volume (PNGV, 1995). This, in turn, requires that they meet all emission and safety requirements and include all of the characteristics that result in strong customer appeal. In addition to product redesign the program embraced research aimed at improving the effectiveness of the broad manufacturing enterprise, including everything from the development of new analytical tools to the use of new materials and manufacturing processes.

The PNGV declaration of intent requires an independent peer review "to comment on the technologies selected for research and progress made." In response to a request in July 1994 by the undersecretary for technology adminis-

[1]USCAR, which predated the formation of PNGV, was established to support intercompany, precompetitive cooperation to reduce the cost of redundant R&D in the face of international competition. Chrysler Corporation merged with Daimler Benz in 1998 to form DaimlerChrysler. USCAR is currently composed of a number of consortia, as shown in Appendix D.

tration, U.S. Department of Commerce, on behalf of PNGV, the National Research Council established the Standing Committee to Review the Research Program of the Partnership for a New Generation of Vehicles. The committee annually reviews PNGV's research program, advises government and industry participants on the program's progress, and identifies significant barriers to success. This is the seventh review by that committee; the previous studies are documented in six National Research Council reports, which also contain background on the PNGV program and the committee's activities (NRC, 1994, 1996, 1997, 1998, 1999, 2000). Chapter 4, "Program Overview," of the current report contains the committee's broad assessment of what has happened in the PNGV program during the past seven years. It also discusses program issues raised by changes that have occurred during this time and suggests that it may be time for a critical review of the program goals.

The PNGV goals and the basis for all of the National Research Council reviews are articulated in the PNGV Program Plan (PNGV, 1995; The White House, 1993):

> **Goal 1. Significantly improve national competitiveness in manufacturing for future generations of vehicles.** Improve the productivity of the U.S. manufacturing base by significantly upgrading U.S. manufacturing technology, including the adoption of agile and flexible manufacturing and reduction of costs and lead times, while reducing the environmental impact and improving quality.
>
> **Goal 2. Implement commercially viable innovations from ongoing research on conventional vehicles.** Pursue technology advances that can lead to improvements in fuel efficiency and reductions in the emissions of standard vehicle designs, while pursuing advances to maintain safety performance. Research will focus on technologies that reduce the demand for energy from the engine and drivetrain. Throughout the research program the industry has pledged to apply those commercially viable technologies resulting from this research that would be expected to increase significantly vehicle fuel efficiency and improve emissions.
>
> **Goal 3. Develop vehicles to achieve up to three times the fuel efficiency of comparable 1994 family sedans.** Increase vehicle fuel efficiency up to three times that of the average 1994 Concorde/Taurus/Lumina automobiles with equivalent cost of ownership adjusted for economics.[2]

As the committee has noted in previous reports and as noted in a number of other studies, achieving significant improvements in automotive fuel economy and developing competitive advanced automotive technologies and vehicles can

[2]As noted in the PNGV Program Plan (PNGV, 1995), the long-term goal of the PNGV program is to develop vehicles that will deliver up to 80 miles per gallon (mpg) or British thermal unit (BTU) equivalent. If an alternative source of energy is used, such as a diesel-powered vehicle or a fuel cell vehicle powered by methanol or hydrogen, the goal will be up to 80 miles per BTU equivalent of a gallon of gasoline (114,132 BTUs). Where values of mpg are used in this report with options not using gasoline, those mpg values are understood to be miles per equivalent gallon of gasoline. "Fuel consumption," also used in this report as an index of energy use, is the reciprocal of fuel economy.

provide important economic benefits to the nation, improve air quality, improve the nation's balance of payments, and reduce emissions of greenhouse gases to the atmosphere (DOE, 1997; NRC, 1992, 1997, 1998; OTA, 1995; PCAST, 1997; Sissine, 1996). The rise in international oil prices and in U.S. gasoline prices during 2000-2001 has created attention in the media about the cost of driving and fuel economy. Even with these price increases, gasoline prices still remain relatively low in real terms, and U.S. automobile purchasers have little incentive to consider fuel economy as a major factor in their purchase decisions. In addition, the sales of light trucks, especially sport utility vehicles (SUVs), which have lower fuel economy than automobiles, account for close to half of the vehicles purchased for personal use. The lack of market incentives in the United States for buyers to purchase vehicles with high fuel economy has made it difficult to realize public benefits from higher-cost components and designs that improve fuel economy. Nevertheless, Ford Motor Company, General Motors Corporation, and DaimlerChrysler Corporation have announced plans to increase the fuel economy of SUVs significantly by the 2003 to 2004 time period.

The PNGV strategy of striving to develop an affordable automobile with a fuel economy of up to 80 mpg that maintains current performance, size, utility, and cost levels while meeting safety and emissions standards would circumvent the lack of economic incentives for buying automobiles with high fuel economy. If the PNGV strategy were successful, buyers could purchase vehicles with all the desirable consumer attributes, as well as greatly enhanced fuel economy; there would be no reason not to do so. The creation of such a vehicle, as the committee has noted in previous reports, is extremely challenging. This ambitious goal has, however, stimulated rapid development worldwide of essential technologies, verifying the strategic value of the PNGV program. Vehicles meeting Goal 3 requirements and the triple-level fuel economy may not be achieved, but many of the technologies developed in an attempt to meet Goal 3 may enter the market in a variety of production vehicles (Goal 2) (e.g., higher-efficiency engines, hybrid drive trains, and regenerative braking), thus providing a significant impact on vehicle fuel consumption.

PNGV is a partnership aimed at simultaneously meeting both business and societal goals. It joins the extensive R&D resources of the national laboratories and university-based research institutions, together with the vehicle design, manufacturing, and marketing capabilities of both the USCAR partners and suppliers to the automotive industry.[3] Government funding for PNGV is primarily used for the development of long-term, high-risk technologies. Funding by USCAR and industry is generally used for the development of technologies with nearer-term commercial potential, the adaptation of government-laboratory technology devel-

[3]The U.S. Department of Commerce, the U.S. Department of Energy, the U.S. Department of Transportation, the Environmental Protection Agency, and the U.S. Department of Defense are the federal partners in the PNGV program.

opments to automotive applications, and the production of proof-of-concept vehicles. Substantial in-house proprietary R&D programs related to PNGV are also under way at USCAR partner facilities.

PNGV has created industry-government technical teams responsible for R&D on the candidate subsystems, such as fuel cells and four-stroke direct-injection engines. A manufacturing team, an electrical and electronics power-conversion devices team, a materials and structures team, and a systems analysis team are also part of the PNGV organization (NRC, 1996, 1997, 1998, 1999, 2000). Technical oversight and coordination are the responsibilities of a vehicle-engineering team, which provides the technical teams with vehicle system requirements supported by the systems analysis team. A safety working group has been formed since the committee's sixth review.

PROGRAM PROGRESS

At program inception, several milestones were established: a technology selection in 1997; concept vehicles in 2000; and production-prototype vehicles in 2004. The program has made substantial progress since its inception and has met major milestones thus far. At the end of 1997 PNGV made a technology selection based on assessments of system configurations for alternative vehicles. Several technology options—such as gas turbines, Stirling engines, ultracapacitors, and flywheels for energy storage—were eliminated as leading candidates. The 10-year span of the program dictated some of these choices. In its fourth review the committee agreed with PNGV's technology selections (e.g., four-stroke, internal-combustion engines, fuel cells, batteries, power electronics, and structural materials). The four-stroke compression-ignition direct-injection (CIDI) engine was selected as the most likely power plant to enable the fuel economy goal to be met within the program time frame; the fuel cell power plant was retained in the program as a highly promising longer-range technology. After the technology-selection process PNGV was able to concentrate its resources on fewer technologies with the intent of defining, developing, and constructing concept vehicles by 2000 and production prototypes by 2004 (PNGV, 1995).

The second major milestone, the development of concept vehicles, was met in early 2000 and was a major program achievement. Using PNGV-developed technologies and their own in-house proprietary technologies, the USCAR companies each developed separate concept vehicles with fuel economies between 70 and 80 mpg. These were reviewed in the committee's sixth report (NRC, 2000). Early in the program the USCAR partners decided it would be impractical to design and build a joint concept car, a decision the committee supports. The three concept-car designs that resulted from this decision each made its own significant contribution to program goals. Clearly it is even more impractical for the development of production-prototype vehicles to be pursued in the cooperative part of the PNGV program. Proprietary issues require production prototypes to be designed, developed and built independently in a single company.

With the development of the concept vehicles, and as the program approaches the 2004 deadline, the nature of the program is shifting. Greater effort is being expended by the USCAR companies as vehicle development, cost reduction, and overcoming manufacturing barriers become increasingly important. In addition, when the technologies under development become more mature and allow improvements in performance and reliability, along with reduced costs, they can be transferred into broader applications in the vehicle fleets.

THE EVOLVING CONTEXT

As with all programs, the context in which the PNGV program was first conceived has evolved, and public policy issues and the economic environment have changed. In assessing progress, reviewing activities, and making judgments about the future success of the PNGV program, it is important to keep a number of issues and changes in mind, some of which have been noted in previous committee reports:

- The power train with the highest probability of meeting the vehicle fuel-economy target of 80 mpg by 2004 is the hybrid-electric power train powered by a CIDI engine. In 1999 approximately midway through the program, the Environmental Protection Agency promulgated Tier 2 emission standards for particulate matter and oxides of nitrogen (NO_x) substantially more stringent than those at the start of the program (Federal Register, 1999). This action brought into question the possibility of meeting these emission requirements with a CIDI engine in a production prototype by 2004. Consequently, a major portion of the program resources was reallocated to address this new development risk. Alternative power plants (e.g., homogeneous spark-ignition engines or gasoline-fueled direct-injection engines) with a higher probability of meeting the Tier 2 standards in the PNGV 2004 time frame would result in vehicles with reduced fuel economy compared with the CIDI engine.
- The market for personal vehicles has changed substantially during the past decade with the sale of light trucks (e.g., pickup trucks, vans, and SUVs) now accounting for about 50 percent of the vehicles purchased for personal transportation. Thus, the traditional family sedan that is part of PNGV's Goal 3 has become a less important component of the "automotive" market. These light trucks are heavier and bigger than automobiles and have lower fuel economy than typical sedans.
- From all the evidence the committee has seen during past reviews, the cost premium of a PNGV-type vehicle with a fuel economy close to 80

mpg will likely be several thousand dollars more than a competing conventional vehicle. Without some form of incentive to overcome this increment, it is unlikely that such a vehicle would be cost-competitive in the market. Thus, the committee recommended during its last review that PNGV should direct its program toward an appropriate compromise between fuel economy and cost using the best available technology to ensure that a market-acceptable production-prototype vehicle that meets emission standards can be achieved by 2004.

- Concerns about the potential for global warming from emissions of greenhouse gases, such as carbon dioxide (CO_2), have become more acute. This has resulted in voluntary agreements by the automakers in Europe and the creation of Japanese weight-class standards for fuel economy aimed at the reduction in CO_2 emissions in the 2008-2010 time frame (Plotkin, 2001).
- Toyota and Honda have introduced hybrid-electric vehicles, the Prius and Insight, respectively, into the marketplace. They are much smaller than the vehicles under development in the PNGV program. Ford, General Motors, and Chrysler have all announced their intention to build some form of hybrid-electric vehicles for production by 2004.
- The PNGV also has pursued the development of fuel-cell-powered vehicles. Much progress in fuel cell technology has been made since the program started, and private companies worldwide are now devoting substantial resources to this effort.
- And finally, in 2001 the Administration has proposed and Congress is discussing an energy policy for the country. As of July 2001, it is too early to determine how this activity will affect the PNGV program.

SCOPE OF REVIEW

The 16-member committee (see Appendix A) that conducted this seventh review of PNGV was tasked as follows:

> The previous six reviews of the PNGV program critically assessed research progress and commented on a number of issues related to the efficacy of the program to meet its goals within the PNGV time frame. In continuation of this independent evaluation, in the Phase 7 review, the NRC Standing Committee to Review the Research Program of the Partnership for a New Generation of Vehicles will address the following tasks:
>
> 1. In light of the PNGV program technical goals and previous NRC Standing Committee recommendations, examine and comment on the overall balance and adequacy of the PNGV research effort, the rate of progress, and the technical objectives and schedules for each of the major technology areas (i.e., fuel cells, 4-stroke direct-injection engines and emissions control, power electronics & electrical systems [electric drive], energy storage, and structural materials).
>
> 2. In light of the emission requirements for the 2004-2010 time period, examine and comment on the ongoing fuels, propulsion engine, and emission-control research efforts to identify and develop commercially viable very-low-emission (e.g., California LEV-2

> and EPA Tier 2 standards) propulsion systems based on fuel cells and 4-stroke direct-injection engines.
>
> 3. Based on its own knowledge of worldwide developments of the technologies under development in the PNGV program, comment on whether PNGV is pursuing appropriate directions to overcome technical barriers. If, in the opinion of the committee, PNGV should be pursuing other approaches based on work ongoing outside of PNGV, the committee will recommend suitable approaches to be taken.

The conclusions and recommendations in this report are based on the committee's meetings, presentations, and other data-gathering activities (see Appendix C). USCAR presented some material as proprietary information under an agreement signed by the National Academy of Sciences, USCAR, and the U.S. Department of Commerce (on behalf of the federal government).

2

Development of Vehicle Subsystems

CANDIDATE SYSTEMS

The success of the PNGV program depends on integrating R&D programs that can collectively improve the fuel efficiency of automobiles within the very stringent boundary conditions of size, reliability, durability, safety, and affordability of today's vehicles. At the same time the vehicles must meet emission regulations, be largely recyclable, and use components that can be mass produced and maintained similar to current automotive products.

To achieve the Goal 3 fuel economy target of 80 mpg (1.25 gallons per 100 miles), the energy conversion efficiency of the chemical conversion system (e.g., a power plant, such as a compression-ignition direct-injection [CIDI] engine or a fuel cell) averaged over a driving cycle will have to be at least 40 percent. This challenging goal requires developing and integrating many vehicle system concepts. For example, the primary power plant will have to be integrated with energy-storage devices and the vehicle structure will have to be built of lightweight materials. Every aspect and function of the vehicle will have to be optimized, both individually and as part of the vehicle system. No aspect of the vehicle function can be left untouched, from minimizing the energy expenditure for maintaining comfort in the passenger cabin to significantly improving the conversion efficiency of the exhaust-gas after-treatment systems.

The USCAR partners have chosen the hybrid-electric vehicle (HEV) as the power train in their respective concept cars. The HEV uses stored energy in the battery to drive an electric motor that provides power boost to the engine, thereby permitting a smaller engine that can be operated closer to optimum conditions. This results in increased energy conversion efficiency, reduced emissions, and the potential to recover a fraction of the vehicle's kinetic energy during braking. Not only were the concept cars demonstrated by the PNGV partners in 2000 a great technological achievement, they also helped clarify the remaining hurdles to achieving success for the PNGV program. It is apparent that the fuel cell will not be feasible in a production-prototype vehicle by 2004. This leaves the internal combustion engine as the primary energy converter, and even using the most efficient one, the CIDI diesel engine, the three-times fuel economy target remains a stretch goal. If maximization of fuel economy is the design target, the diesel engine is the first choice; however, meeting the mandated emission standards is a major challenge for the diesel engine. Therefore, a critical consideration to maximize fuel economy is reduction of nitrogen oxides (NO_x) and particulates, the two emission standards that are most difficult for the diesel engine to meet.

Reducing the cost for manufacturing and moving the concept technology into current and future vehicles has now become a central factor, so there has been a notable shift in emphasis during 2000-2001 to cost reduction and manufacturability. Resources are being focused on continued development of enabling technologies, such as exhaust-gas after-treatment systems, fuel composition effects on system performance, advanced battery energy storage systems, and power electronics and component cost reduction. The investigations of promising longer-term prospects, such as advanced combustion systems and fuel cell technologies, are also continuing.

The PNGV presented to the committee an overview of the status and critical development issues of the candidate energy-conversion and energy-storage technologies that survived the 1997 technology selection process. Overviews of candidate electrical and electronic systems and advanced structural materials for the vehicle body were also presented.

This chapter addresses the following technology areas and related issues:

- Four-stroke internal-combustion reciprocating engines;
- Fuel cells;
- Electrochemical storage systems (rechargeable batteries);
- Power electronics and electrical systems;
- Structural materials;
- Vehicle safety; and
- Fuels.

The committee reviewed R&D programs for each of these technologies, along with related vehicle safety and fuels issues, to assess progress so far and the developments required for the future. In the committee's opinion the PNGV continues to make significant progress in developing the candidate systems and identifying critical technologies that must be addressed to make each system viable. The committee is pleased that, since the introduction of the concept vehicles in 2000, there has been a shift in program focus to include affordability.

INTERNAL COMBUSTION RECIPROCATING ENGINES

The internal combustion engine continues to be the primary candidate power plant for meeting near-term PNGV program goals. To meet the fuel economy target (80 mpg) of Goal 3 the internal combustion engine will have to be integrated into an HEV configuration. The CIDI engine, using diesel fuel, is the most efficient of the internal combustion engines. Consequently, in the near term, taking advantage of the high efficiency of the diesel engine and integrating it into an HEV is the most promising way to attain maximum vehicle fuel economy. However, the challenges of meeting the new California Air Resources Board (CARB) and the U.S. Environmental Protection Agency Tier 2 emission standards are a major hurdle for the CIDI engine (NRC, 1999, 2000), even when used in an HEV power train. In maintaining its quest for a vehicle with fuel consumption of 1.25 gallons per 100 miles (80 mpg) the PNGV has continued its focus on the diesel engine as the primary energy converter for the vehicle. As a result, PNGV has continued its aggressive investigation of different approaches to emission reduction for CIDI engines. To achieve the emission targets will require an integrated approach to further refinements in engine design and operation, aggressive development of exhaust-gas after-treatment and its integration into the power-train system, and modification of the fuel to allow optimum engine performance while facilitating the exhaust-gas emission-reduction technologies. The most notable fuel modification is the need to reduce the sulfur content in the fuel. These three aspects of reducing NO_x and particulate matter were discussed in the committee's sixth report (NRC, 2000). The partnership continued its emphasis on the system approach during 2000-2001. The primary emphasis continues to be exhaust-gas after-treatment systems; however, work continues on fundamental combustion systems, such as homogeneous charge compression ignition (HCCI), basic injection, and combustion analysis, that if successful could offer some emissions reduction and fuel economy improvement.

There is a general sentiment that emission standards can be met with incremental development of known technology applied to homogeneous-charge spark-ignition engines and probably with direct-injection spark-ignited gasoline engines. Using the spark-ignited engine, however, would result in a reduction in the vehicle's fuel economy compared with that of a diesel engine. Each of the partners has proprietary in-house programs on both homogeneous and direct-injection spark-ignition engines. The fuel economy of gasoline direct-injection engines and the challenge of reducing their emissions to meet CARB low-emission vehicle (LEV 2) and federal Tier 2 standards falls between those of the homogeneous spark-ignited gasoline engine and the diesel engine. Because the in-house programs are proprietary, those efforts are not reported here. This review focuses on the status of the partners' joint R&D efforts for diesel engines and the identification of the critical barriers that need to be surmounted for success in the PNGV program.

Program Status and Plans

The unveiling of the concept cars by each of the three USCAR partners in 2000 represented a transition in the PNGV program. The concept cars are a successful technology demonstration that also serves to focus more sharply the critical technical hurdles remaining for successful completion of the program. Clearly, manufacturability and affordability are two critical issues on which PNGV has now placed increased emphasis. The post-concept-car aspect of the program also signifies a shift to a higher degree of in-house proprietary effort by each of the partners as plans are made to transfer the concept car technologies into their respective products. The committee was given individual proprietary briefings by each of the partners to help it understand the extent of these efforts.

In addition to the proprietary internal work PNGV continued its programs of collaborative work on precompetitive fundamentals. This involves work on the interaction between fuel composition and engine performance, investigations of combustion fundamentals and diagnostics, and emission control systems, primarily exhaust-gas after-treatment. The partnership's activity in the four-stroke direct- injection (4SDI) engine technical area is a major collaborative effort of the individual partners, the national laboratories, and a few universities. The Pacific Northwest National Laboratory, Lawrence Livermore National Laboratory, Lawrence Berkeley National Laboratory, Sandia National Laboratories, Los Alamos National Laboratory, National Renewable Energy Laboratory, Argonne National Laboratory, Oak Ridge National Laboratory, and the Department of Energy Headquarters are all active collaborators with the partnership. University participation includes Wayne State University, the University of Wisconsin, and the University of Michigan.

It is important to remember that development of a vehicle power train that maximizes fuel economy while minimizing emissions requires a power-train systems approach. There will be interactions among the fuel, the engine, and the exhaust-gas after-treatment subsystems. PNGV is addressing this systems issue; however, for the purposes of the following discussion, it is convenient to divide it into three components: engine combustion, emission control systems, and engine-fuel interactions.

Engine-Combustion System Developments

The challenge of meeting the CARB LEV 2 and Tier 2 emission standards with a diesel engine was highlighted in the 2000 committee report (NRC, 2000). A critical requirement is that tailpipe NO_x and particulate matter (PM) emissions will need to be drastically reduced for the diesel engine to become a viable PNGV power plant. It is unlikely that requisite emission reductions will be achieved through in-cylinder combustion modification alone; exhaust-gas after-treatment will almost certainly be necessary. However, a reduction in engine-out emissions reduces the after-treatment conversion efficiency required. Furthermore, it may be necessary to tailor the engine exhaust composition to the after-treatment device, which means being able to control the combustion process. Therefore, understanding combustion fundamentals continues to be important.

The engine combustion investigations reported to the committee this year had as objectives: further understanding of in-cylinder combustion phenomena, development of diagnostics that would facilitate optimization of in-cylinder combustion, and demonstration of a fuel system for dimethyl ether, which is a non-sooting fuel.

The partnership has established a draft technical R&D plan and engine technology performance targets for the 80 mpg (1.25 gallons per 100 miles) PNGV vehicle. The Technical R&D Plan Draft Objectives for CIDI engines operating on diesel fuel are (Howden, 2000):

- By 2002 develop and validate NO_x (0.2 g/mile) and PM (0.02 g/mile) emission control technologies;
- By 2004 develop and validate NO_x (0.07 g/mile) and PM (0.01 g/mile) emission control technologies; and
- By 2007 demonstrate the capability to meet Tier 2 Bin 3 emissions (NO_x: 0.03 g/mile; PM: 0.01 g/mile).[1]

The engine technical targets are given in Table 2-1. The emission targets in the table are for the tailpipe output, considering the combined system of the fuel, engine, and after-treatment devices. The 4SDI technical team also quantified emission targets for the engine alone: 0.36 g/mile for NO_x and 0.04 g/mile for PM. The difference between these levels and the 2004 targets of 0.07 and 0.01 g/mile for NO_x and PM will need to be achieved through the exhaust-gas after-treatment system.

It is convenient to categorize the projects in the 4SDI program as those that address short-term issues and those that address longer-term issues. In the engine combustion subprogram, the near-term projects include (1) developing advanced fuel injection systems; (2) studying the cylinder-to-cylinder distribution and transient response of exhaust-gas recirculation (EGR); (3) developing in-cylinder combustion and PM measurements; and (4) performing detailed comparisons between combustion data obtained in an optical research engine, a similar metal engine, and the predictions of sophisticated three-dimensional computer simulations. Projects in the longer time frame focus on continued investigation of HCCI

[1]The California LEV 2 targets for NO_x and particulates are LEV NO_x = 0.05 g/mile at 50,000 miles and LEV PM = 0.01 g/mile at 100,000 miles. This is simplified for illustrative purposes; the California emission regulations are a complex set of rules.

TABLE 2-1 CIDI Engine Technology R&D Plan Technical Targets for an 80-mpg PNGV Vehicle

Characteristics	Year		
	2002	2004	2007
Best brake thermal efficiency (%)[a]	44	45	46
Best full load thermal efficiency (%)[b]	42	43	44
Displacement power density (kW/L)[c]	42	45	47
Specific power (W/kg)	590	625	650
Durability (hours)	>3,500	>5,000	>5,000
Emission control cost ($/kW)[d]	5	4	3
Exhaust emission control device volume (L/L)[e]	3	2	1.5
Engine cost ($/kW)[d,f]	30	30	30
NO_x emissions[g] (g/mile)[h]	0.2	0.07	0.03
PM emissions[g] (g/mile)[i]	0.02	0.01	0.01
Fuel economy penalty due to emission control system (%)	<5	<5	<5

[a]Ratio of mechanical power out to fuel energy rate (lower heating value) in.
[b]Ratio of mechanical power out to fuel energy rate (lower heating value) in at peak power.
[c]Ratio of the peak power output divided by the volumetric displacement of the engine.
[d]Assumed high-volume production of 500,000 units per year.
[e]Volume of emission control system (in liters) per liter of engine displacement.
[f]Constant out-year cost targets reflect the objective of maintaining engine system cost while increasing engine complexity.
[g]Full-useful-life emissions using advanced petroleum-based fuels as measured over the federal test procedure as used for certification in those years.
[h]NO_x values given are tailpipe emissions; the engine-out target for NO_x is 0.36 g/mile. The difference between this number and the tailpipe emission target will need to be achieved through after-treatment systems.
[i]PM values given are tailpipe emissions; the engine-out target for PM is 0.04 g/mile. The difference between this number and the tailpipe emission target will need to be achieved through after-treatment systems.

combustion, with assessments of how advanced variable compression ratio engine concepts and electromagnetically or hydraulically actuated engine valves—often referred to as camless engines—could enhance combustion control.

It is generally agreed that EGR will have to be employed to meet emission standards. Typical engines tend to exhibit a quadratic increase of PM emissions with EGR increases. Furthermore, the engine's tolerance for EGR varies inversely with load. The exact reasons for these behaviors are not known. Through its fundamental research programs PNGV is developing a diagnostic for use in production engines for measuring the cylinder-to-cylinder EGR distribution. Being able to minimize or eliminate the maldistribution of EGR between different engine cylinders should enable maximum exploitation of this emission reduction technology.

More insight into the fuel-injection and air-entrainment processes was gained through optical diagnostics in a high-pressure, high-temperature constant-volume vessel. For a further understanding of in-cylinder particulate phenomena a collaborative effort was started between PNGV and Sandia National Laboratories to address real-time particulate diagnostics for size, number density, and volume fraction. Real-time measurements of the particulate emissions during a cold start have been made for both a turbocharged diesel-powered vehicle and one with a port-injected spark-ignition engine. Even though the measurements are qualitative, the results of the particulate measurements for relative size distributions and mass concentrations for the two engines are consistent with those obtained using conventional measurement techniques during steady-state operation. The particulate levels from the spark-ignition engine are approximately one order of magnitude below those of the diesel engine. The size data are important, as there is growing concern about a correlation between particulate size and adverse health effects. If consensus is reached on the importance of the particulate size and adverse health effects, the partnership will be in a position to assess the status of the engine it is developing.

The collaborative effort between Sandia National Laboratories, Wayne State University, and the University of Wisconsin, Madison, continues. In this program an optical engine, metal engine, and advanced computer simulation are being used to enhance PNGV's fundamental understanding of in-cylinder processes. The combination of optical diagnostics, metal engine operation, and computational comparison are being used to improve the capabilities of the simulation, which in turn is being implemented for design optimization.

A more basic approach to in-cylinder emission reduction is being pursued through PNGV's program on HCCI combustion. PNGV's research and an exhaustive literature review have confirmed that HCCI combustion is limited to light-load operation. Therefore, before HCCI combustion can be employed in an automotive engine, methods of control must be developed. Commonly used methods to control HCCI combustion include variable valve timing, variable compression ratio, variable residual gas retention, active ignition via hot surfaces or auxiliary injection, variable mixture preparation schemes, and variable fuel chemistry. The PNGV is active in evaluating the potential of these techniques through combustion diagnostics (being done at Sandia National Laboratories) and fuel composition-engine interactions, and investigations of variable valve actuation and variable-compression-ratio systems. The challenges are many and great; however, if successfully developed, these combustion approaches could be integrated into a more conventional engine-operating scheme to improve emissions and fuel economy, but probably not for the 2004 time frame.

In 2000, it was reported that the 4SDI technical team, working with the Department of Energy, awarded Cummins Engine Company and Detroit Diesel

Corporation cooperative agreements to develop PNGV-size CIDI engines. Since the announcement the respective companies have formed alliances with catalyst manufacturers and are developing complete CIDI engine systems, including exhaust-gas after-treatment.

After-Treatment and Controls and Sensors

The most difficult challenge facing the 4SDI technical team is to develop power trains that meet the legally mandated emission standards. At this point in the program the only hope for meeting these standards is through the development of very effective exhaust-gas after-treatment systems. As was the case during the committee's sixth review, the 4SDI technical team devoted a majority of its effort to this activity in 2001. The primary focus of its effort is reducing the NO_x and PM emissions from the diesel engine.

NO_x After-Treatment Systems. The 4SDI technical team presented its strategy for NO_x control technology development to the committee. The plan is to pursue extensive in-cylinder EGR, which will require optimizing the EGR systems in conjunction with NO_x-reduction after-treatment systems. After-treatment techniques being studied include selective catalytic reduction (SCR) systems using urea (NH_2CONH_2); active lean-NO_x reduction systems using fuel as the reductant; NO_x absorbers or traps with catalyst systems; and nonthermal-plasma catalytic systems. Passive lean-NO_x catalytic systems have been dropped from consideration. The 4SDI technical team believes that the major breakthrough in conversion efficiency needed to make them viable is unlikely in program's time frame.

The highest NO_x conversion efficiency and the least fuel economy penalty have been achieved using an SCR system with urea as the reductant. Conversion efficiencies of approximately 80 percent over the Federal Test Procedure (FTP) driving cycle have been obtained with simulated feed gas. This resulted in a projected fuel economy penalty of less than 0.5 percent. Engine tests using a 1.2-liter DIATA (direct-injection, aluminum, through-bolt assembly) engine indicated 70 percent NO_x conversion efficiency. The remaining technical issues for SCR with urea are avoiding production of ammonium nitrate and ammonium sulfate under low-temperature operation; eliminating ammonia slip at high temperatures; size and cost of the system; developing a viable way to store, deliver, and replace the urea onboard the vehicle; and infrastructure issues for urea distribution and delivery to the vehicle. A urea SCR system will also require tailpipe NO_x and NH_3 sensors for closed-loop control of urea injection.

Active lean-NO_x catalyst systems use hydrocarbon injection, most likely fuel, upstream of the catalyst to promote the NO_x reduction. The best performance of such systems to date yields conversion efficiencies of less that 50 percent, with fuel economy penalties of 5 to 8 percent. This efficiency is too low to meet emission standards. In addition, the temperature range for peak conversion is too narrow and occurs at too low a temperature to match the maximum efficiency operating range of the engine. In addition, the catalysts are susceptible to moisture degradation. For active lean-NO_x catalyst technology to be viable, the conversion efficiencies, moisture resistance, and light-off temperature range will all have to be improved (i.e., new catalysts must be developed).

Nitrogen oxide traps have been demonstrated to work very well. However, they are sensitive to sulfur contamination. This sulfur sensitivity does not appear to have a minimum threshold. Any sulfur in the exhaust will contaminate the trap. Reducing the sulfur level in the fuel but not eliminating it will only reduce the rate at which the trap becomes contaminated. Test data indicated that even with a fuel sulfur level of 3 ppm, the absorber's NO_x-trapping efficiency was reduced from 95 percent to 80 percent in just 250 hours. Regeneration of the trap to get rid of the sulfur requires temperatures on the order of 650 to 700°C for up to 10 minutes and adversely affects the trap's durability. Even if the sulfur contamination problem can be eliminated, the NO_x traps, like the active catalyst systems, require regeneration by injection of hydrocarbons, again most likely fuel. Consequently there is a fuel economy penalty associated with NO_x traps. The 4SDI technical team estimates that this fuel economy penalty would be approximately 5 percent.

Adding a sulfur trap would extend the lifetime of current NO_x traps. This means that either a sulfur trap regeneration procedure must be developed or a routine maintenance schedule for sulfur trap replacement must be mandated. Successful development of a viable sulfur trap would then pave the way for use of other sulfur-sensitive technologies (e.g., continuously regenerating particulate traps). To address this critical need DOE's Office of Advanced Automotive Technologies (OAAT), working with the PNGV, has included sulfur-trap development and sulfur-tolerant catalyst material development in a recent solicitation. The request for proposals details specifications for the sulfur trap and the sulfur-tolerant catalysts that are directly aimed at PNGV needs. The funding levels are $1.5 million, with a 20 percent industry cost share, for each topic. The program duration is two years.

Although it is still in the research stage, there is growing optimism about the potential of nonthermal plasma catalysis for NO_x, and possibly particulate, control (much of the work is being conducted at the Pacific Northwest National Laboratory). Creating a plasma within a dielectric catalytic material enhances NO_x reduction. It is believed that the plasma converts some of the NO in the exhaust gas to NO_2 and partially oxidizes the hydrocarbon reductant to intermediate hydrocarbon species, aldehydes, and carbon monoxide (CO). These intermediate hydrocarbon species and the NO_2 are then involved in the reduction reactions at the NO_x catalyst. An oxidation catalyst is then needed downstream of the plasma catalyst to remove the residual hydrocarbons (HC) and CO. To date the best performance achieved in the laboratory has been a NO_x conversion of approximately 60 percent with a fuel economy penalty of approximately 5 percent. (A conversion efficiency of greater than 95 percent will be required to meet Bin 2 of the federal Tier 2 emission regulations.) Development work is continuing to investigate different catalyst combinations and is pursuing techniques for reducing the energy consumption. One interesting experiment was performed that applied the plasma system to a urea SCR system while simulating engine cold start. The temperature at the end of the test was only 140°C, and excellent low-temperature performance was achieved.

One of the reasons for the high level of interest in plasma catalysis is that the NO_2 generated by the plasma as part of the NO_x reduction process is also known to be an oxidant for PM. It may be possible to incorporate a plasma catalysis system that will reduce both NO_x and particulates in the same device. An experimental plasma-catalyst diesel particulate filter was tested. The catalyst bed served as a trap for the larger particles and seemed to oxidize the smaller ones. It successfully reduced the particulate mass emissions over a particle size range of 10 to 350 nm. When the device was tested without the plasma operating, the passage of particles smaller than 200 nm became excessive, peaking at 700 million particles per cubic centimeter at a particle size of approximately 50 nm. When the device was operated with the plasma, the particle flux was on the order of 1 million particles per cubic centimeter for all size ranges. This is approximately the particle emission rate for a homogeneous spark-ignited engine. Chemical analysis of the particulates indicated that the plasma was also effective in reducing the polycyclic aromatic content of the particulate matter. Such characteristics of the particulate matter are not currently subject to regulation; however, these data are important as they establish a database for evaluation relative to the findings of current investigations into the possible link between particulate composition and size and adverse health effects.

The above results are encouraging. However, the catalyst bed to which the plasma is applied eventually became clogged with particulate matter and needed to be regenerated, a common problem for all diesel particulate filters. In total, the plasma-catalyst after-treatment technology is an exciting program. The plasma catalyst system offers the advantages of NO_x removal over a wide temperature range, low NO_2 production, and very-low-temperature NO_x removal with urea SCR and possibly even PM removal. Because it is a new technology, there is reason for optimism that it will improve with further maturity. The significant challenges at this time are its electrical power consumption and the question of whether future improvements in conversion efficiency will be sufficient to meet future emission standards, which will undoubtedly be lower than the current CARB LEV 2 and Tier 2 standards.

Particulate Control Systems. PM reduction strategies include fuel injection optimization, regenerative particulate traps, and fuel and lubricant modifications, which will be discussed below in the "Engine-Fuel Interactions" section. The degradation of fuel economy, as well as the expense, is a concern for all these approaches. Most vehicles will require particulate traps to meet the new PM emission standards. The two major issues for the particulate traps are how well they filter out very small particles (d < 250 nm) and how easily they can be regenerated. The 4SDI technical team is pursuing particulate trap regeneration technology and is closely following the activity in Europe, where in 2001 Peugeot is planning to introduce a regenerative particulate trap that uses the Rhodia Eolys™ cerium-based fuel additive.

In addition to the cerium-based fuel additive, techniques for particulate trap regeneration include microwave ignition, electrical ignition, and catalytic regeneration with a catalyst in the trap (continually regenerating trap [CRT]). The CRT oxidizes the PM using NO_2, which is produced catalytically at the entrance of the trap. The catalyst to produce NO_2 is sulfur sensitive, and consequently the CRT requires low-sulfur fuel or a sulfur trap. The PNGV is actively investigating or evaluating all these approaches to trap regeneration.

Although particle size is not a component of the emission standards, increased attention is being given to the size distribution of the PM emitted from combustion systems. Of concern is the complex relationship between particle size, particle number, and total particulate matter mass and adverse health effects with a suggestion that particle number may be especially important for ultrafine particles less than 500 nm in diameter. The 4SDI technical team has started to build such a database. The particle sizes measured range from approximately 50 to 500 nm. Tests were conducted for steady-state operation with a filter but without regeneration. The results indicated that the particle size distributions before and after the filter were similar in terms of the number of particles as a function of size. However, the number density and volume fraction of particulates were reduced by over 99 percent.

The assessment of the filters as a PM control technology remains optimistic. The primary issues that must be addressed are the durability, regeneration efficiency, and cost of these devices.

Vehicle Testing. In an effort to move closer to practical demonstration of the different after-treatment systems and to address the issues of integrating them into the entire power-train system, the 4SDI technical team is performing integrated power-train vehicle testing. Two different research vehicles fitted with experimental after-treatment systems, particulate traps, and a NO_x absorber or urea SCR system are being tested on a special low-sulfur fuel (4 ppm sulfur). Results from these tests indicate that under ideal conditions Tier 2 emission standards are achievable with small diesel vehicles and advanced controls. Of course, these results are from highly controlled research tests. The life of the NO_x absorber and the extent to which it is affected by sulfur in the exhaust remain problematic. Reducing sulfur sensitivity, enhancing absorber performance, and developing improved regeneration will be the focus of further efforts. Much work is still required.

The 4SDI engine still faces major challenges: NH_3 slip,[2] ammonium nitrate generation, urea introduction, and catalyst activity for urea SCR systems; degradation from sulfur of NO_x traps and continually regenerating particulate traps; system fuel economy reduction; and system sensing and control. It also appears that, if NO_x traps are to be used, a method of regeneration to remove the sulfur must be developed. As discussed above, this is a very challenging problem. If a breakthrough is not achieved in the next 18 to 24 months, it is unlikely that the diesel engine will be able to achieve the production-prototype status by 2004.

Sensors and Controls. PNGV and DOE's OAAT have recognized that precise and interactive control systems will be necessary to achieve optimization of the power train, combining the fuel, engine, and after-treatment system. Critical to developing these control systems is the availability of sensors to supply the requisite inputs to execute the control strategy. Therefore, sensor development is included in a recent OAAT solicitation. The request for proposals describes a program of $1.5 million for three years with a 25 percent industry cost share. New sensors for NO_x, PM, and O_2 are requested. The performance specifications for these sensors directly address PNGV needs. The requested deliverables are laboratory bench-level demonstrations after 12 months. Based on the results of the bench tests, a decision will be made regarding continuation. The end goal of the program is engine dynamometer demonstrations of the respective sensors.

Engine-Fuel Interactions

The interactions among fuel, engine, and exhaust-gas after-treatment system are very complex and full of multi-faceted trade-offs. The cost of producing the fuel, modifications required to the existing infrastructure, and the trade-off between engine efficiency and emissions all must be considered. If new fuel performance specifications are necessary, most likely the government will have to be involved to oversee their introduction into the market. By necessity, the energy industry (i.e., the fuel companies), relevant government agencies, and the PNGV partners need to collaborate in the program.

In an effort continued from the previous year, the investigation of the effects of fuel chemistry and physical properties on engine performance and emissions was an area of intense activity this year. Partnership efforts included auto and oil company ad hoc test programs; the Advanced Petroleum-Based Fuel–Diesel Emission Control (APBF-DEC) program; an ultra-clean fuels initiative; a Coordinating Research Council Advanced Vehicle/Fuel/Lubricants committee; the CARB fuel cell fuel program; and EUCAR (European Council for Automotive Research and Development)/USCAR cooperative fuels research programs.

---

[2]Ammonia slip refers to the emission of unconverted ammonia or ammonium compounds (sulfate and nitrate) in the exhaust.

The ad hoc test program between the oil companies and the PNGV partners represents a continuation of the effort reported in the committee's sixth report (NRC, 2000). The objective of the program is to identify advanced diesel fuels and fuel properties that would enable the successful use of compression ignition engines to meet the new emission standards. The energy companies participating in the ad hoc program are BP-Amoco, ExxonMobil, Shell, Marathon-Ashland, Citgo, and Equilon. In the program each of the USCAR partners tested four fuels in their own PNGV CIDI engines. In addition, DOE has evaluated each of the fuels in a commercial light-duty CIDI engine. The fuels tested were a CARB commercial #2 diesel as the baseline, a low-sulfur, low-aromatics, high-cetane petroleum-based diesel (LSHC), a neat Fischer-Tropsch diesel (FT-100), and a blend of 15 percent dimethoxymethane (DMM15) and 85 percent LSHC. These fuels were chosen to determine whether fuel composition might have significant effects on exhaust emissions, and no attempt was made at this stage to conform to all of the fuel specifications, such as whether they can be used in winter climates. A more detailed listing of the fuel properties and the different engine operating conditions under which they were tested is given in Table 2-2 and Table 2-3.

During this year the scope of the ad hoc program was enlarged. Several new objectives were added. First, attempts were made to assess the contribution of lubricating oil to PM and NO_x emissions. Second, a sharper focus on identifying the desirable characteristics of fuel oxygenates was undertaken. Finally, a more detailed chemical characterization of the PM was incorporated into the data analysis procedure.

The initial phase of this work concentrated on minimizing the engine-out PM and NO_x emissions using steady-state dynamometer tests. The next phase of the program will concentrate on tailpipe emissions involving after-treatment systems, including transient tests, and a different set of fuels. The results reported to the committee were only for the first phase of the program. The PNGV made the results public through publication in technical papers, namely the Society of Automotive Engineers (SAE) Congress 2001 (Gardner et al., 2001; Hilden et al., 2001; Kenny et al., 2001; Korn, 2001; Szymkowski et al., 2001).

To date, test results indicate that the fuel does have an effect on the engine-out PM and NO_x emissions; however, there was significant variation in the effectiveness of the fuel for reducing emissions for different operating conditions and from engine to engine. In some cases the emissions were worse for the test fuels than for the CARB base fuel. In general the Fischer-Tropsch and DMM/LSHC fuels exhibited the greatest reduction in particulate matter. At some conditions a factor-of-two reduction in PM was observed relative to the CARB base fuel. Nitrogen oxide emissions were not so dramatically affected. The greatest reduction in NO_x seen with the test fuels relative to the base CARB fuel was 30 percent. This is somewhat encouraging; however, the level of reduction needed to meet the Tier 2 standards is closer to an order of magnitude, as opposed to a factor of two. Thus, none of the fuels tested would have enabled the engine to meet the

TABLE 2-2 Fuel Properties of Ad Hoc Fuel Test Program

Fuel Property	CARB[a]	LSHC[b]	DMM15[c]	FT-100[d]
Specific gravity, 15°C	0.8379	0.8168	0.8208	0.7803
Initial boiling point, °C	191	208	41	222
T10, °C	215	232	64	257
T50, °C	253	277	262	288
T90, °C	308	321	317	324
End point, °C	330	343	341	337
Cetane number	48.4	64.6	59.4	81.1
Mono aromatics, wt%	15.0	8.9	8.4	1.2
PAH[e] (Di and Tri), wt%	5.1	1.0	0.8	0.0
Total aromatics, wt%	20.1	9.9	9.2	1.2
Carbon (C), wt%	86.48	85.70	80.92	84.77
Hydrogen (H), wt%	13.48	14.30	13.70	15.12
Oxygen (O), wt%	0.05	0.00	5.38	0.11
Nitrogen (N), ppmw	9	<1	<1	2
Sulfur (S), ppmw	175	1	<2	0
Flash point, °C	72	85	<2	98
Cloud point, °C	–24	–4	–6	0
Pour point, °C	–33	–7	–9	–1
Kinematic viscosity (40°C), cSt	2.457	2.921	1.861	3.204
HHV[f] kJ/g	45.9	46.3	43.2	47.2
LHV[g] kJ/g	42.6	43.3	40.8	43.9

[a]CARB = CARB commercial #2 diesel fuel.
[b]LSHC = low-sulfur, low-aromatics, high-cetane petroleum-based diesel.
[c]DMM15 = blend of 15 percent dimethoxymethane and 85 percent LSHC.
[d]FT-100 = neat Fischer-Tropsch diesel fuel.
[e]PAH = polyaromatic hydrocarbon.
[f]HHV = higher heating value.
[g]LHV = lower heating value.

emission standards without extensive after-treatment. Consequently, Phase 2 of the program, in which the fuels are tested in a power-train system that includes after-treatment, is very important. It was also found that the lubricating oil could contribute from 0 to 36 percent of the particulate mass. The contribution depends on the engine-operating mode. Light-load operation results in the largest lubricating oil contribution.

Because of the encouraging results of PM reduction from oxygenated fuels, 71 potential oxygenate blending agents were studied. The oxygenates were evalu-

TABLE 2-3 Engine Operating Conditions of Ad Hoc Fuel Test Program

Engine Speed (rpm) and Brake Mean Effective Pressure (bar)	Moderate EGR Schedule (%)	Time Weighting Factors (seconds)	
		GM	DaimlerChrysler/Ford
Idle, 1,200 rpm/0.1 bar	40	700	N/A
1,500 rpm/2.62 bar	30	600	600
2,000 rpm/2.0 bar	30	375	375
2,300 rpm/4.2 bar	18	200	200
2,600 rpm/8.8 bar[a]	7	25	25

[a]DaimlerChrysler tests were run at 2,500 rpm at this condition.

ated on the basis of their oxygen content, flash point, solubility, stability, viscosity, cetane number, lubricity, elastomer compatibility, potential toxicity, biodegradability, and air quality impact. From the 71 candidate compounds 2 were chosen for future tests: di-butyl maleate and tripropylene glycol monomethyl ether.

Phase 1 of the ad hoc program is now complete. Discussions are under way for the complete emission-control testing systems evaluation, Phase 2. The projected timetable for Phase 2 is on the order of two years.

Since the APBF-DEC program is just getting under way, no technical results are available for evaluation. The mission of the APBF-DEC program is to identify optimal combinations of fuels, lubricants, diesel engines, and emission-control systems to meet continually decreasing emission standards, while maintaining customer satisfaction. Attention also is being given to the possibility of additional emission constraints, such as unregulated substances and ultrafine particulate matter. It is estimated that $35 million will be needed for the program. A government and industry steering committee and working groups will guide the program.

As the PNGV plans its technical program for the future, knowledge of the characteristics of the fuel that will be available in the market and the timetable for introduction of "new" fuels is critical. The partnership feels there will be little change in commercial fuel quality before 2004. Perhaps there will be incremental decreases in the fuel sulfur level and some public test programs of low-sulfur fuel in this time frame. In the 2004 to 2008 time frame, the sulfur level of highway diesel fuel will drop to a 15-ppm cap at the pump. This level of sulfur could enable advanced emission control systems to be introduced. Beyond 2008 advanced fuels will be needed. It is not clear exactly what their performance characteristics should be; however, it is critical that the joint industry-government research programs actively participate in determining what characteristics are needed and provide a feasible way to bring these new fuels to the market.

Assessment

The 4SDI technical team has made excellent progress in continuing the development of a power-train system to meet the PNGV goals. The year 2000 USCAR concept cars are a triumph in integrating promising technologies into demonstration vehicles. As it should be, activities are now concentrating on transferring the technologies demonstrated in the concept cars to the products of the respective partners. These activities are in progress in each of the USCAR companies.

In the 4SDI technical program the challenge continues to be addressing the trade-off between fuel economy and emission reduction technologies. It is generally believed that the spark-ignition engine can meet the emission standards through further development of existing emission reduction technologies. However, the spark-ignition engine, either homogeneous charge or direct-injection gasoline, does not offer the fuel economy improvement potential of the CIDI engine. Each of the partners has proprietary in-house programs addressing the continued improvement of spark-ignition engines. In the pre-competitive cooperative programs, the diesel engine continues to be the focus of research as the desired power plant for the PNGV program. It offers the potential for the best fuel economy with the most realizable near-term manufacturability. The critical issue for the diesel engine continues to be whether the emission standards for NO_x and particulate matter can be met. At this point in the program the prospect of meeting the emission targets with the CIDI engine is improving but is still speculative.

Even though significant progress has been made in the area of exhaust-gas after-treatment and its integration into an engine power train, there is still no clear "winning" technology that has emerged. Each technology has its attractions and its deficiencies. It appears that particulate traps as a generic approach have emerged as the most viable approach for particulate reduction; however, the method of regenerating the traps is still an open issue. After-treatment reduction of NO_x is still a wide-open arena, with many subtle and difficult technical issues for each approach under consideration. NO_x traps work well but suffer from extreme sulfur sensitivity. Urea SCR systems are sulfur insensitive but suffer from issues of how to incorporate urea into the vehicle and support for a new infrastructure. Plasma reduction systems are promising but still very much in the research stage. All after-treatment systems will introduce new sensing and system control challenges that are just now beginning to be addressed. In summary, there are no clear winners and it is not yet certain whether sufficient development can be done between now and 2004 to enable the diesel engine, with an after-treatment system, to meet the requisite emission standards. The next 18 to 24 months will be critical.

The sulfur content of highway diesel fuel will be reduced to a 15-ppm cap by a recently promulgated EPA regulation, but it is not known whether this level will be an enabler for the emission reduction technologies under consideration. The fuel composition will most likely be a factor, but the best composition and properties have yet to be determined. Nor is it likely that fuel composition changes will eliminate or drastically reduce the required conversion efficiencies of the after-treatment systems. Critical issues have been identified and programs are addressing them. The activity is intense, but the time is short.

Global competitiveness is one of the objectives of the PNGV program. The statement of task asks that the committee comment on PNGV activities in light of the committee's knowledge of worldwide R&D on the various technologies under development for advanced, high-fuel-economy vehicles (see Chapter 1). As part of its activities the committee solicited an overview presentation on engine and after-treatment development in Europe from a representative of FEV, a respected international company in this area (see Appendix C). This gave the committee one benchmark on the technology status of the PNGV partnership. The topic, engine and after-treatment development, was chosen because it is one of the most critical technical areas in which PNGV is working, and is in essence a critical decision point for determining whether the maximum fuel economy of the vehicles under development in the PNGV will be met. Based on this presentation, and its own knowledge of developments occurring worldwide, the committee believes that the partnership is indeed at the cutting edge in terms of its knowledge and understanding of light-duty diesel engine technologies and exhaust-gas after-treatment development.

The committee also believes, however, that, because the European community is actively developing, manufacturing, and marketing diesel-powered light-duty vehicles, most new diesel engine developments will probably emanate from Europe, rather than from PNGV. Assessments of the technical status of the partnership in the areas of homogeneous-charge and direct-injection, spark-ignited engines are more difficult; these are highly proprietary areas for the companies. There is some sentiment that the Japanese may be the technical leaders in these areas, especially the development of the direct-injection, spark-ignited engines.

Recommendation

Recommendation. The PNGV should continue the aggressive pursuit and development of lean-combustion exhaust-gas after-treatment systems. The PNGV should also work to develop a detailed systems-modeling effort to quantify the fuel economy penalty associated with using different technologies to meet the emission standards. These efforts should include quantification of the extent to which vehicle hybridization can be used to reduce emissions and the fuel consumption impact of changing the vehicle's primary energy converter.

FUEL CELLS

Even though the fuel cell portion of the PNGV has been accepted as being on a longer time scale than other candidate technologies, it still represents an advantageous and potentially viable alternative. No other energy converter appears to have a better potential for combined low-emission, high-energy conversion efficiency than fuel cells. There are, though, many substantial barriers remaining to the realization of a mass-manufactured consumer vehicle. These barriers include performance as well as physical, fuel-related, and cost issues.

Resolution of issues concerning component physical properties, system arrangement and behavior, performance, and cost has continued to progress through the PNGV efforts, and no insurmountable barriers have been identified. On the other hand, progress is much slower than expected (based on original targets and schedules), and some of the development results to date are still far short of those needed for automotive manufacturing viability. As development shifts more toward automobile integration concerns, other types of issues, such as those affecting drivability (start-up time, transient capability) and those associated with a range of operating conditions and hostile environments (long exposure to low temperatures, desert conditions, climbing long hills or mountains), must be resolved. In addition, of course, many mass-manufacturing issues are far from resolution, and many are undoubtedly still unknown.

Some of the problem areas can be mitigated in the near term by the use of gaseous hydrogen as the onboard stored fuel. This fuel choice eases (but does not really solve) cost problems and virtually eliminates most drivability concerns. This is because fuel processors used to provide hydrogen from stored onboard methanol, gasoline, or other hydrogen-bearing fuels add cost and time delays (as well as weight and volume) compared to a system using onboard stored gaseous hydrogen.

It has been recognized for some time that, for fuel-cell-powered vehicles, fuel selection is not an independent issue but is very much tied to the successes in onboard fuel processor technology development. Furthermore, the onboard fuel energy conversion efficiency is also very much affected by fuel choice. An onboard "gasoline" fuel processor, for example, can reduce energy conversion efficiency as much as 10 to 15 percentage points, thus reducing (or even eliminating) the efficiency advantage of a fuel cell over an internal combustion engine. Indeed, the onboard use of pure hydrogen fuel results in the least complex system, as well as the highest onboard fuel energy conversion efficiency; however, hydrogen is currently produced almost entirely from natural gas in a process that also involves considerable energy loss and the generation of emissions, including CO_2. There are additional energy losses and emissions produced in compressing and distributing the hydrogen for ultimate use in fuel cell vehicles. Hydrogen is, in addition, more expensive per unit of fuel energy than gasoline and, due to onboard storage volume limitations, typically results in a reduced vehicle range compared with liquid hydrocarbon fuels. Also significant with regard to the use of hydrogen as a vehicle fuel is the almost complete lack of required infrastructure. A comprehensive analysis of overall "well to wheels" efficiency and cost for various fuels and propulsion systems, including hydrogen and fuel cells, has recently been published (Weiss et al., 2000).

The reduced complexity and cost, as well as the improved drivability, will likely ensure that the first generation of fuel-cell-powered automobiles will be fueled by hydrogen. That hydrogen is a likely short-term fuel choice was further evidenced by the 2000 Tokyo Automobile Show, where most major manufacturers displayed fuel-cell-powered concept cars and all but the DaimlerChrysler Necar 5 and the Jeep Commander were hydrogen fueled. The Necar 5 and Jeep Commander were methanol fueled and were significant in that the passenger compartment of neither, including the small A-class vehicle, was compromised by the fuel cell or fuel-storage components. With pressurized hydrogen, most of the vehicles had diminished passenger compartments or trunk space, primarily to accommodate the volume needed for hydrogen storage. A few of the modified production vehicles, like the International Fuel Cells (IFC) hydrogen-fueled fuel cell system integrated into the Hyundai Santa Fe sport utility vehicle (SUV), showed little or no reduction of passenger or cargo space (e.g., all components, including the hydrogen tanks, were in the engine compartment or under the floor pan [ground clearance was reduced a bit]), but it is doubtful whether any of the fuel cell vehicles (except perhaps methanol-fueled vehicles) would provide a range in excess of 100 miles.

The first generation of fuel cell vehicles is expected, based on projections from the major automobile manufacturers, to be produced by the 2003-2005 time frame. If the manufacturers select (as seems likely) pressurized hydrogen as the fuel of choice, then these first vehicles will almost certainly be limited to very narrow and select markets (such as certain fleet vehicle applications) where range and fuel infrastructure are not major considerations. The availability of fuel cell vehicles to the general public will be delayed until an acceptable fuel infrastructure can be provided for a "new" fuel or until an acceptable fuel-cell-powered vehicle can be manufactured that can use the existing fuel infrastructure. Most of the PNGV development efforts are oriented toward the latter of the two scenarios, although there are also efforts directed toward the former. This is felt to be an appropriate distribution of effort, since at least for the foreseeable future, both vehicle range and consumer fuel cost favor the use of liquid fuels and the existing fuel infrastructure.

The PNGV fuel cell technology development program involves two distinctly different types of activities: those oriented toward achieving certain levels of fuel cell performance (efficiency, emissions, life) and those oriented primarily toward compatibility with the manufacture and marketing of automobiles (size, weight, noise, start-up time). Cost is a factor for virtually all fuel cell applications, but is clearly more of an issue for automotive than for stationary applications. Thus, many of the targets for the PNGV fuel cell programs are far more stringent than they would be for nonautomotive applications. The significance of this is that, even though industry fuel cell developers appear to have reached a self-sustaining level of activity, the adaptation of fuel cells to automobiles would probably be on a much longer time scale without the PNGV efforts.

Program Status

The Year 2000 and Targets

Early in the PNGV program the year 2000 was chosen as a major milestone year for fuel cell technology development. The year 2004 was selected as the final milestone in the development program, but the year 2000 represented a time when intermediate targets were scheduled to have been met. Consequently, it was also a year that required an in-depth assessment of the development program and an evaluation of both the appropriateness of the individual targets and actual progress made toward meeting each target. The year 2000 targets covered a range of efficiency, emissions, physical, and cost parameters and were focused on approaching the composite of attributes needed for a gasoline-fueled[3] fuel cell energy converter ready to be seriously considered as an automobile power plant manufacturing alternative. "Approaching" is a key word in this context since the year 2004 was chosen as the year actually to meet these composite attributes. Thus the 2000 targets were somewhat less demanding than the 2004 targets.

The year 2004 was already an extension of the original PNGV Goal 3 technology development schedule. This extension was made several years ago for fuel cells when it became clear that (1) the original time schedule projections were not realistic due to the very immature status of fuel cell development and (2) fuel cell systems should not be dropped from the program but retained as a long-term component since there were (and are) many potential benefits to society associated with the technology. Extending the schedule for fuel cell development did not involve compromising the ultimate targets, which represented huge (in some cases orders of magnitude) advances in the technologies when compared with the then current state of technology.

As the fuel cell development programs continued into the year 2000 it again became clear that, in spite of impressive progress in virtually every area of activity, intermediate targets were not going to be met. This reality also carried the implication that the ultimate goals were also unlikely to be achieved by the year 2004. However, as before, the substantial progress that had been made and the continuing recognition of the potential benefits, especially in the area of

[3]Gasoline in this context refers to a petroleum-based fuel similar to present gasoline that could use the existing infrastructure.

extremely low emissions and high energy efficiency, when operating on hydrogen, suggested a continuation with further modified targets and/or schedules. Also relevant to this issue are the individual targets that were established years ago and represented combined judgments about what was necessary, and also what might be accomplished, for a realistic fuel-cell-powered automobile. These judgments were based on adapting fuel cells to then existing concepts of advanced vehicles, and without access to some of the modeling and simulation tools that are currently available to the developers. In addition, this committee recommended in its sixth report that "PNGV should conduct trade-off analyses to establish relative priorities for fuel-cell technical targets and cost targets" (NRC, 2000). The basis for the committee's recommendation was that, since it was already clear that not all targets were going to be met, the PNGV technical team should establish which targets were the most critical to be met (or even tightened) and which could be loosened without compromising the long-term PNGV goal of a marketable consumer fuel cell automobile. As a result, the targets and schedules were revisited by the Fuel Cell Technical Team members, who are recommending essentially a four-year extension with most of the year 2000 targets shifting to 2004 and the 2004 targets shifting to 2008. The Fuel Cell Technical Team also recommended a few changes in the magnitude of individual targets.

The Year 2000 Status and Progress Toward Targets

As reported by the Fuel Cell Technical Team, the only year 2000 target that was achieved for the complete integrated gasoline fuel cell system was in the area of emissions; however, just operating the integrated gasoline fuel cell system successfully in 2000 must be considered an important event. Prior to 2000 the gasoline fuel processor and the stack sub-system were developed separately and then operated together, but controlled separately, to demonstrate the capability to operate on gasoline reformate (and other hydrocarbon fuels). Indeed, the lack of an operational integrated gasoline system by late 1999 was noted as a major concern by this committee. It was recognized that many of the important system issues could not be known, much less resolved, until a truly integrated system became operational.

Fortunately, in the year 2000, two integrated gasoline systems became operational:

1. The system composed of a 50-kW Nuvera fuel processor and Plug Power stack (and balance of plant) and
2. An IFC 50-kW fuel processor/stack/balance of plant.

The Nuvera/Plug Power system, and projections based on this system, are the basis for most of the current status and progress toward targets reported by the Fuel Cell Technical Team.

There has been significant progress toward all year 2000 targets, although none except emissions (below Tier 2 levels) has been demonstrated or projected as being able to be met from limited testing to date for the gasoline system. Specific power and power density, for example, are projected to be about 140 W/kg and 140 W/L, compared with year 2000 targets of 250 and year 2004 targets of 300 for both. This means that, based on current technology projections, the complete gasoline fuel cell system would be roughly twice as large and twice as heavy as targeted for the 2004 vehicle. While this is still an area of concern, the scope of progress is evident when compared with similar parameters early in the PNGV, which were at least 10 times the target values. Similar trends are noted with cost values, which have been reduced from projected values of several thousands of dollars per kilowatt to about $300/kW. Like the previous parameters, even with the progress, cost is a major concern when the $300/kW is compared with the 2000 target of $150/kW and the 2004 target of $50/kW.

Start-up time has been reduced from tens of minutes to about six minutes currently, compared with 2000 and 2004 targets of one minute and one-half minute, respectively. Again, there has been substantial progress, but it is still an area of concern. The same is true in such areas as response time and durability, where progress has been good but targets remain elusive. Overall system fuel efficiency at 25 percent of peak power (which is where most operation is expected to occur) is in the mid-30s (in percent) where the 2000 and 2004 targets are 40 percent and 48 percent, respectively. The current efficiency is excellent compared with the *average* efficiency of about 20 percent obtained with present-day spark-ignited automobile engines. However, the year 2000 PNGV concept cars of Ford, GM, and DaimlerChrysler are HEVs with turbocharged diesel engines. Hybrid-electric vehicle engines operate much closer to peak efficiencies on average than do nonhybrid vehicle engines and probably yield average engine efficiencies also in the mid-30s (in percent) and perhaps even a bit higher.

It is encouraging that there has been a steady increase in the efficiency of fuel processors as the fuel cell development programs have continued. Thus, while current values are high enough that efficiency is not necessarily a *major* concern, it is clear that continued progress toward the target values for "gasoline" fuel cell systems should be a high priority if such fuel cells are to offer a clear benefit in fuel efficiency over competitive internal-combustion-engine HEV systems.

Significant Accomplishments

There have been significant accomplishments in essentially all developmental areas. For the most part, these were evolutionary advances that resulted in systems, sub-systems, and components being smaller, lighter, less costly, better performing, and more durable. Among these are:

- A smaller 50-kW fuel-flexible fuel processor (Nuvera);

- A more sulfur-tolerant reformer catalyst (ANL);
- An air-stable, nonprecious-metal catalyst for fuel processing (ANL);
- A high-power-density reformate-capable stack (Honeywell);
- Improved cathode catalysts (LANL); and
- A lightweight composite bipolar plate (ORNL).

In addition, there were some essentially new products or processes worth noting, such as:

- A 12-kW microchannel steam reformer (PNNL);
- Carbon-foam heat exchangers (ORNL);
- A high-volume fabrication technique for membrane electrode assemblies (3M); and
- A direct-methanol fuel cell for portable power (LANL).

There were major accomplishments in successfully marrying two fuel-flexible fuel processors with two reformate-capable stacks into two self-contained integrated systems, one by Plug Power/Nuvera and one by IFC. Only limited information has been obtained regarding these two integrated systems, but it appears that both are successfully operating as self-contained units. The Plug Power/Nuvera system is pressurized while the IFC is near ambient pressure. Continued operation of these systems will help clarify the actual (versus perceived) benefits of the two approaches. Also, it should be noted that even though the technology is still in a much earlier developmental stage than the proton-exchange-membrane (PEM) fuel cell, the direct-methanol technology is advancing rapidly. If sufficiently developed, it has the advantage of using a liquid fuel without a fuel processor, thus making it very attractive for automotive applications. At present it still uses large quantities of platinum catalyst and operates with relatively low efficiency, but it continues to show progress.

While details are not being made public, it is known that there is considerable foreign effort in automotive fuel cell technologies, especially in Japan and Germany. In addition to its obvious integrated fuel cell vehicle work such as the methanol-fueled FCEV experimental vehicle introduced about three years ago, Toyota claims to have a new (and better) hydrogen-absorbing material to store hydrogen onboard a vehicle. A significantly improved hydrogen storage material that is low-cost, safe, and long lived would be one major step toward the feasibility of a hydrogen-fueled consumer vehicle. DaimlerChrysler in Germany has successfully integrated a methanol-fueled fuel cell energy converter with fuel processor into the small A-Class vehicle without infringing on passenger or storage space. This Necar 5 was shown publicly in early 2001 and is now undergoing testing.

In summary, the PNGV fuel cell developmental efforts seem to be well organized and are focusing on the more important areas of concern. In spite of good progress, 2000 targets have not been met and the Fuel Cell Technical Team is recommending essentially a four-year extension in the targets (see Table 2-4). Considering the continued outlook for potential advantages of fuel cells, this seems to be a reasonable recommendation.

Recommendations

Recommendation. Because of the potential for near-zero tailpipe emissions and high energy efficiency of the fuel cell, the PNGV should continue research and development efforts on fuel cells even though achievement of performance and cost targets will have to be extended substantially beyond original expectations.

Recommendation. To help establish target priorities, the PNGV should continue evaluation of the relative importance of various fuel cell development targets through trade-off and sensitivity studies.

Recommendation. The PNGV should use the two 50-kW integrated gasoline fuel cell energy converters (Nuvera/Plug Power and International Fuel Cells) to the extent feasible to understand and characterize further pressurized versus ambient-pressure fuel cell systems.

ELECTROCHEMICAL ENERGY STORAGE

The PNGV program to develop electrochemical energy storage technology for HEVs has focused on advanced batteries—primarily the nickel metal hydride (NiMH) and lithium-ion (Li-ion) electrochemical systems—for about the last seven years. High-power design versions of these batteries were thought to hold the best prospects for meeting the stringent performance, life, and cost targets established early in the PNGV program for the energy storage subsystem of fully competitive HEVs. The soundness of choosing these systems for development is confirmed by the substantial progress made by PNGV toward most of these targets and the commercial use by all Japanese HEVs of either NiMH or Li-ion batteries.

The committee's sixth report addressed the PNGV targets for energy storage systems and reviewed NiMH and Li-ion technology development against these targets, with the general conclusions that, despite significant progress, calendar life, cost, and safety remained concerns for Li-ion technology, which is receiving the bulk of PNGV's battery R&D funds (NRC, 2000). Nickel metal hydride HEV batteries have not quite met performance targets, and, as with Li-ion batteries, projected costs have exceeded targets by about a factor of three. The sixth report noted the leadership of Japanese battery manufacturers, especially in NiMH high-power-battery development and commercial applications in HEVs. These appli-

TABLE 2-4 Proposed Revised Technical Targets for Integrated Fuel Cell Power Systems[a]

			Calendar Year		Old Targets	
Characteristics	Units	Status	2004	2008	2000	2004
Energy efficiency @ 25% of peak power	%	34	40	44	40	48
Energy efficiency @ peak power	%	31	33	35	None	None
Power density	W_e/L	140	250	325	250	300
Specific power	W_e/kg	140	250	325	250	300
Cost	$/k$W_e$	300	125	45	130	50
Transient response (10 to 90% power)	sec	15	5	1	3	1
Cold start-up (–20°C to maximum power)	min	10	2	1	2	1
Cold start-up (20°C to maximum power)	min	<5	<1	<0.5	1	0.05
Survivability	°C	–20	–30	–40	None	None
Emissions	—	<Tier2Bin2	<Tier2Bin2	<Tier2Bin2	Tier 2	Tier 2
Durability	hours	1,000	4,000	5,000	2,000	5,000

[a]Includes fuel processor, stack, auxiliaries, and start-up devices (excludes gasoline tank and vehicle traction electronics).

cations appear to be technically successful and are beginning to create markets for HEVs and batteries even though the Japanese HEV batteries have not yet demonstrated PNGV target life, and costs are more than three times higher than PNGV target costs.

Finally, the sixth report recommended that the PNGV energy storage targets be refined to be fully consistent with the evolving models for optimization of HEV drive trains and operations, and that the advanced materials being developed to overcome Li-ion life, safety, and cost issues be applied in, and validated through, the technologies emerging from the program's main battery development contractors. Together with well-supported development targets, these observations and recommendations are considered an appropriate framework for the current assessment of the PNGV battery program.

Program Status and Progress

Presentations made to the committee at its December 7-8, 2000, meeting, as part of its review of the PNGV program and PNGV's responses to the committee's follow-on questions, showed again that the battery program is well organized in terms of task structure, milestones, and the technical review process. The program is overseen by PNGV's Electrochemical Energy Storage (EES) Technical Team, which represents diverse talents and can draw on the extensive knowledge and resources of the participating automobile manufacturers and federal organizations, primarily DOE and its national laboratories.

The program has met key milestones with the delivery of a full-scale, complete Li-ion HEV battery (SAFT) and 50-V Li-ion battery modules (Polystor). Other significant indicators of progress include calendar life improvements of the program's Li-ion technologies, higher-power NiMH electrode assemblies (GMO), a lower-cost NiMH module design (VARTA), and the initial transfer of advanced Li-ion electrode active materials from ANL to several of the battery developers. The EES Technical Team has revised key battery characterization and test procedures (especially with respect to the dual-mode HEV application) to serve as more realistic representations of anticipated operating conditions. The EES Technical Team also increased the battery life targets as noted in Table 2-5.

The supporting basic research at the national laboratories has contributed important insights on Li-ion cell failure mechanisms and the cause of thermal runaway. Cell and module testing has increased in importance and effort, and it is benefiting from the program's independent battery testing capability at the Idaho National Environmental and Engineering Laboratory. The agreements with the Japanese LIBES (Lithium Ion Battery Energy Storage) program and with EUCAR to discuss calendar-life testing methods also are positive steps. For the first time the HEV batteries being tested by PNGV include Japanese prototypical Li-ion batteries acquired from Shin-Kobe, a subsidiary of Hitachi and leading developer of Li-ion electric vehicle and HEV batteries.

TABLE 2-5 New Targets for Batteries

	Power-Assist HEV Battery[a] Targets		Dual-Mode HEV Battery[a] Targets	
	Previous	New	Previous	New
Shallow cycles (25 Wh)	200 k	300 k	n.a.[b]	n.a.[b]
Dual-mode cycles (80% depth of discharge)	—	—	2,500	3,750
Calendar life (years)	10	15	10	15

[a]In the power-assist mode, the battery is used only briefly (e.g., up to 20 seconds at any time) to assist the primary power source in the acceleration process. The target of 300 Wh available energy permits such a battery to deliver sufficient energy for several accelerations.

In the dual-use mode the battery provides propulsion energy beyond the acceleration process because of the inherently slow response of the primary power source, such as a fuel cell during start-up. Depending on the time and power profile over which the battery needs to deliver propulsion energy, such a battery needs to have an available energy and storage capacity several times larger than a power-assist battery.

[b]n.a. = not applicable.

All the NiMH and Li-ion technologies developed and evaluated in the program are expected to meet the key technical performance targets for the power-assist HEV application: peak pulse power of 25 kW discharge for 18 s, peak pulse recharge/regenerative power of 30 kW for 2 s, and 300 Wh available energy at peak pulse power, all from a nominally 40 kg battery pack. (Although pack weights exceeded 40 kg in a number of tests, so did the measured performance parameters. Prorating battery weight to 40 kg would still leave Li-ion performance well above targets and NiMH so close that evolutionary design improvements should attain the targets.) The dual-mode battery designs similarly are meeting performance targets, in that case with batteries weighing significantly less than the 100-kg target. The summary charts presented to the committee include data on the Delphi Lithium Ion Polymer (LIP) and Argotech/Avestor lithium (metal) polymer battery technologies recently added to the PNGV program. These advanced batteries, too, meet the performance targets for the power-assist HEV application; in the case of Avestor's lithium polymer battery, performance targets are also met for the dual-mode HEV application.

Achievement of sufficiently long battery calendar life continues to be a challenge, especially for Li-ion cells and batteries, but even for the inherently more stable NiMH battery chemistry, achievement of the earlier 10-year-life target has not been proven. The data presented indicate promising life for both NiMH and Li-ion batteries in the power-assist and dual-mode designs and test cycles; however, no actual cycle- or calendar-life test data were presented to the committee. In response to a committee request the EES Technical Team provided some data on elevated-temperature capacity retention of SAFT Li-ion "standard" cells that suggest a significant life improvement of 1999 over 1998 cell technology; however, these life tests were accelerated through elevation of temperature and extended over less than one and a half years. It is questionable whether the seven-year Li-ion battery calendar life projected from these data is realistic, since it apparently is not supported by cell- and battery-level life models verified through real-time testing. This reservation applies to most or all of the calendar-life data obtained in the program to date.

Attaining the PNGV unit cost targets of $300 for power-assist batteries and $500 for dual-mode batteries for any of the program's technologies remains the most difficult program challenge. The volume production costs projected for current-technology power-assist batteries exceed the target by a factor of four to five for Li-ion batteries and by a factor of two to three for NiMH batteries, with similarly large ratios for dual-mode designs. Appropriately, these discrepancies drive much of the program's R&D strategy and efforts, including the emphasis on lower-cost designs, materials, and manufacturing techniques for NiMH and Li-ion batteries, and the pursuit of additional advanced battery technologies (Li-ion polymer; Li [metal] polymer). Although the program is making some progress toward these broad goals, no breakthroughs with reasonable prospects for the dramatic cost reductions needed to meet targets have been achieved.

Assessment of the Program

Several dimensions are relevant when assessing PNGV's electrochemical energy storage and battery program. First, the core of the program consists of well-organized and technically managed development activities centered on the most promising battery systems at several leading developers. Several testing, diagnostic, and materials development projects are supporting core development efforts.

These efforts are guided by performance, life, and cost targets that were developed early in the program. These targets are based on the postulate that HEV performance and cost have to be competitive with those of the corresponding internal-combustion-engine-powered vehicle, but their derivation from the targets for, and characteristics of, an optimized HEV propulsion system has not yet been fully explained to the committee.

As noted above, the battery-cycle and calendar-life targets, originally set to assure a 10-year battery life, have now been raised to 15 years, nominally the life of the vehicle. While this rationale seems reasonable on the surface, in the committee's view even the previous 10-year life target is ambitious (especially for Li-ion and other lithium-based batteries) because of the electrochemical materials transformations and chemical corrosion processes occurring in such batteries. To date no rechargeable battery of any type has shown realistic potential for achieving a 15-year life in high-power cycling service during which temperatures of 40°C or more are likely to be experienced by the battery. The 15-year-life target thus could presage the nominal failure of the program's battery development efforts. Looking ahead, the need to provide a correspondingly long warranty could well discourage prospective manufacturers from making the investments needed to commercialize such batteries.

Similarly, the realism of the cost targets set for power-assist HEV batteries and, even more so, dual-mode HEV batteries may be questioned from two standpoints. First, the probability of attaining such low costs with batteries also meeting life targets must be considered low because inherently less expensive (larger-scale and lower-power) electric vehicle versions of NiMH and Li-ion batteries are unlikely to cost less than $300/kWh in volume production (Anderman et al., 2000). Even cost-optimized 1.5-kWh Li-ion or NiMH batteries are, therefore, unlikely to cost less than $500-$750 per unit in future mass production. Reducing these costs through development of advanced, less expensive materials or lower-cost manufacturing techniques is an important PNGV program goal. However, more than one materials cost breakthrough would be needed for the NiMH or Li-ion technologies, and no such breakthroughs are apparent at this time. Reducing battery cost by reducing capacity (while still demanding that 300 Wh be available at peak pulse power) will result in higher-cost designs and shorter cycle life due to the greater cycling depth and the associated stresses on the battery. Consistent with this view, the projections of the PNGV battery developers for current technology are substantially higher than $500, as are informally obtained estimates of the costs of the batteries used in the currently commercial Japanese HEVs.

The cost issue is even more serious for dual-mode HEV batteries because the specified available energy of 1.5 kWh strongly suggests that the nominal capacity of a dual-mode HEV battery needs to be three to five times larger than the nominal capacity of a power-assist battery. Even allowing for savings due to their lower-power design, dual-mode batteries are likely to cost at least $1,000 to $1,500 per battery unit, or $670-$1,000 per kWh of available energy.

Thus, although most technical aspects of PNGV's battery development program have been progressing satisfactorily, prospects for reaching the newly defined 15-year life targets and the cost targets for both power-assist and dual-mode HEV batteries would appear slim, even with continued, significant improvements of the mainstream Li-ion and NiMH materials and manufacturing technologies. Prospects for the newly added Li-ion polymer battery are unlikely to be better since the battery uses similar materials and probably will have higher manufacturing costs due to the need to produce the required thin-film polymeric electrolytes and integrate them into cells. The cost projection for the lithium (metal) polymer HEV battery was even higher, which led PNGV to drop this technology rather quickly, and in the committee's view, correctly, from the program.

Given that neither the battery life nor the cost targets are likely to be attained, the committee believes that it is appropriate at this stage of the program to examine critically whether these targets are still appropriate, and how the cost and competitiveness prospects of HEVs are likely to be affected by various levels of battery life and costs that fall short of targets.

In past reports the committee has pointed to the leadership position of Japanese battery companies in NiMH and Li-ion batteries for HEV applications. The PNGV battery program is working on the same battery systems and has achieved comparable performance although the Japanese NiMH HEV battery technology appears to have superior life capability and is ahead in battery manufacturing. If the PNGV's ongoing efforts to develop new, significantly lower cost materials for NiMH and Li-ion HEV batteries are successful, this could result in a technology-based cost leadership position.

One intention of the PNGV program is to help provide viable battery technology choices for the HEV development and commercialization efforts. PNGV's contributions to these choices thus are an important measure when assessing the PNGV battery program and its impacts. In this context the committee notes that the pre-prototype HEV batteries used in the PNGV concept vehicles use battery technologies developed in the PNGV program or, in one case (ArgoTech LIP battery), in the United States Advanced Battery Consortium (USABC) program overseen by the same technical team. The PNGV program connection is less direct for some of the batteries selected by the same manufacturers for their first-generation HEVs intended for commercialization, presumably because they had to rely on technologies that, while not meeting all PNGV targets, are likely to be commercially available in the near term. This points to changing opportunities and therefore an evolving role of the PNGV battery program: to develop technologies that represent major advances—especially longer life and lower cost—over the best commercial or near-commercial NiMH and Li-ion HEV batteries. Success in sustained efforts to achieve such advances would not only enable increasingly viable HEVs but could help establish battery technology leadership as a basis for market competitiveness for the program's battery industry partners. Seen in that context, the current battery life and cost targets remain meaningful, but as stretch goals rather than criteria for success or failure of the PNGV battery program.

Recommendations

Recommendation. PNGV should conduct an independent study of Li-ion battery life predictions and prediction models for a realistic assessment of Li-ion life capability. The assessment should include calendar life and cycle life, and it should be used to assess the prospects of Li-ion batteries to meet the new life targets, with emphasis on the 15-year calendar life.

Recommendation. Working with the major developers and independent battery cost experts, PNGV should produce best-case cost projections or estimates for Li-ion and NiMH batteries in mass production, based on the materials cost reductions that might be feasible without and with breakthroughs. PNGV should develop a systematic hybrid-electric vehicle component cost trade-off database that allows the impact of above-target costs of batteries to be determined.

Recommendation. On the basis of the information obtained in the life and cost assessments and trade-offs recommended above, PNGV should critically examine and appropriately revise hybrid-electric vehicle (HEV) battery life and cost targets, retaining the current targets as stretch goals for future HEV battery technologies and the R&D needed to achieve the necessary breakthroughs. To these targets, PNGV should add targets that, if met, make the program's battery technologies viable choices for limited, nearer-term applications, similar to the mid-term targets adopted by the United States Advanced Battery Consortium for battery electric vehicles.

POWER ELECTRONICS AND ELECTRICAL SYSTEMS

All the advanced vehicles being developed under the PNGV program are variants of an HEV. They therefore incorporate some degree of electric propulsion, the energy for which comes from a high-voltage battery. The power-electronic interface between the battery and the motor provides the necessary energy management and electric traction control. The electrical accessories (e.g., ventilating fans, lights, entertainment systems) will require a low-voltage supply, typically 12 V. A 12-V battery charged from the high-voltage battery through a power-electronic converter will most likely provide this supply. These power electronic, electrical accessory, and motor subsystems and their interconnections are the subject of the Electrical and Electronics Systems Technical Team's (EE Tech Team's) work.

In the last committee report, the significant influence of the electrical and electronic systems on total vehicle cost was highlighted (NRC, 2000). At that time it was reported that significant technical innovations in component and manufacturing technologies were needed to meet the aggressive PNGV 2004 cost targets. These cost targets and current values are shown in Table 2-6. The EE Tech Team is still facing a major challenge to reduce these costs. The efficiency targets are equally aggressive and require a reduction of 50 percent in the losses for both the power electronics and motor and generator subsystems.

Program Status and Progress

The EE Tech Team is addressing the cost challenge with three targeted programs: (1) develop an integrated power module specifically for the automo-

TABLE 2-6 Current Status and Targets for Power Electronics and Motors

Component	Specific Power	Efficiency	Cost
Power electronics			
Today	4 kW/kg	95%	$10/kW[a]
2004 target	5 kW/kg	97-98%	$7/kW
Motor/generator			
Today	1.5 kW/kg	92%	$6/kW
2004 target	1.6 kW/kg	96%	$4/kW

[a]Based on proprietary cost models.
SOURCE: PNGV, 1999.

tive application (the automotive integrated power module [AIPM]); (2) develop an automotive electric motor drive (the AEMD); and (3) engage the resources and programs of the national laboratories to develop advanced components for power-electronic systems. During the last year the team has also conducted cost-gap analyses to identify the technology and manufacturing opportunities that are capable of closing the gap between target and present costs. The primary affordability elements are:

- High-volume, low-manufacturing-cost motor designs;
- Thermal-management systems;
- Low-cost, high-performance materials for components, especially capacitors, permanent magnets, and high-voltage connectors;
- The integration of the AIPM with the AEMD;
- Cost of system reliability (15 years/150,000 miles); and
- Cost of the high-current AIPM for the 42-V accessories supply.

The three manufacturers under contract to develop the AIPM—Semicron, Rockwell/Silicon Power Corporation (SPCO), and SatCon—displayed their current hardware for the committee. The technical approach being taken by each was described in last year's committee report (NRC, 2000). Except for one of the contractors, the hardware displayed was not far removed from the conceptual stage. The EE Tech Team has required each contractor to execute a detailed cost-gap analysis in collaboration with suppliers to address four issues:

1. Cost analysis and benchmarking of commercially available products;
2. Identification of technologies for closing the cost, performance, weight, and volume gaps;
3. Selection of partners to supply key technologies; and
4. Development of a schedule for implementation of necessary technologies in the program.

After this analysis all three AIPM contractors expressed confidence that the target of $7/kW could be met. Rockwell shared detailed data with the committee. The data show that the year 2000 cost for the Rockwell design is $17/kW, but Rockwell hopes to reach the target of $7/kW for a 55-kW peak-rated AIPM by 2003. Cost reductions of $438 for materials and $108 in labor and overhead are required to achieve this. Descriptions of the other analyses were not made available to the committee, but the EE Tech Team has expressed confidence in the results. Because of the limited supporting evidence provided, the committee is concerned that these expectations are overly optimistic. To monitor progress toward these targets, the EE Tech Team should require its gap analysis to be updated on a regular and frequent basis.

The AEMD contractors are Lynx/Delco Remy and Delphi. The principal challenge is cost, as the power density of available machines is nearly equal to the 2004 target (see Table 2-4). During the past year the AEMD targets have been modified with the addition of the gearbox. These new targets, which are very challenging, are shown in Table 2-7. Lynx/Delco Remy is pursuing an axial-gap, permanent-magnet design with innovative winding structures and rare-earth permanent magnets. The company's attention has been directed to the motor design and not the issues associated with installing the machine in a vehicle. Delphi's approach is to use a more conventional radial-gap induction machine with a number of cost-reducing design innovations for fabricating the rotor and stator laminations. Delphi has also considered the challenge of integrating the machine in a vehicle, and has produced designs for both series and parallel HEV configurations.

Efficiency improvements are expected to come from replacing the presently used induction machine with a permanent-magnet machine. Work being done at the National Aeronautics and Space Administration's Ames Laboratory on bonded magnetic materials is directed at making the manufacturing process of permanent-magnet machines more economically attractive.

TABLE 2-7 New Target Specifications for Two System Sizes for the AEMD

	New Targets with Gearbox	
	30 kW	53 kW
Cost	$300	$450
Weight	<22 kg	<35 kg
Volume	<7 L	<11 L
Efficiency	>93%	>93%

SOURCE: Provided by PNGV in response to committee questions.

Programs at the national laboratories and universities continue to be leveraged by the EE Tech Team, particularly for the development of component technologies. The carbon-foam, thermal-material work at Oak Ridge National Laboratory (ORNL) has been licensed to Poco Graphite Corporation, thereby guaranteeing a supply of material suitable for evaluation by contractors and others working on advanced power-electronic systems. Argonne National Laboratory and Ames are developing high-energy, permanent-magnet materials suitable for high-density motors. ORNL and Virginia Polytechnic Institute continue their work on circuit topologies and control techniques for advanced converters. ORNL also has a program to design drives for low-inductance machines, such as those being developed by Lynx/Delco Remy under its AEMD contract.

Low-cost, high-dielectric-constant materials for capacitors are critical to the solution of the energy-storage (filter) problem in converters. Sandia National Laboratories has been working on this problem and has developed new materials based on polyconjugated aromatics. Five chemistries have been evaluated, and AVX Corporation is commercializing components based on this new material.

Silicon carbide power semiconductor devices have the potential to reduce switching and conduction losses substantially while operating at considerably higher temperatures than silicon devices, resulting in increased efficiency and reduced cooling requirements and weight. This technology is being developed by a large number of organizations including the Office of Naval Research, Purdue University, DaimlerChrysler Corporation, Cree, and the U.S. Tank Automotive Command. To date the only functionally practical device to be demonstrated has been the Schottky diode. During the last few years, however, the purity of the material has been improved by nearly an order of magnitude and costs have been reduced to approximately $20 for 2-inch wafers. The technology has been licensed to Vishay, and the EE Tech Team has proposed that the DOE fund continued development.

Efficiency improvements in the power electronics are expected to come primarily from the availability of generation-4 insulated-gate bipolar transistors (IGBT), the major semiconductor device in these systems. Current devices have a forward drop on the order of 3 volts. Semiconductor manufacturers are anticipating a drop on the order of 1.5 volts for their generation-4 devices. Additional efficiency improvements are expected from magnetic and dielectric material developments under way at the national laboratories.

The challenge of integrating the AIPM and AEMD is being addressed by Unique Mobility under a Small Business Innovation Research Program contract. Phase 1 of this program has resulted in a demonstrated design with thermal performance significantly superior to conventional designs. Phase 2 of the program is intended to optimize the cooling design and result in a family of modular motor set designs for powers between 15 and 60 kW.

Assessment of the Program

The committee has previously stated that the functional specifications of the power electronics and electrical systems have been met (NRC, 1999). The remaining challenges are meeting the physical (packaging), efficiency, and particularly, cost targets. While progress is being made in all these dimensions, there has yet to be a demonstration of the integration of advances to produce a system that displays significant improvements. This said, however, the committee applauds the energy and rigor with which the EE Tech Team has organized and managed its mission. The gap analysis process in particular is a very powerful tool for building confidence that targets have hope of being met. The committee is impressed with the thoroughness with which the EE Tech Team has searched out, evaluated, and incorporated relevant ongoing work in the government and private sectors into the PNGV program.

In the committee's opinion the AIPM is the most critical element on the path to meeting the EE system targets. Furthermore, capacitor development is essential for the success of the AIPM. The three AIPM contractors are pursuing different manufacturing systems, and there seem to be substantial differences in their progress to date. The gap analyses are excellent planning tools, but the EE Tech Team should require that the contractors provide physical demonstrations of their manufacturing designs, specifically in the form of the AIPM. It is also not clear that the communication between the AIPM, AEMD, and system integration contractors, and the materials and component developers, is as close as it should be to obtain the fastest incorporation of component developments into the system designs and analyses.

Meeting both the cost and efficiency targets requires the development of significant new technology. While progress is being made in this regard, the apparent rate of progress is not sufficient to give the committee a high degree of confidence that these targets will be met by 2004.

Recommendations

Recommendation. The automotive integrated power module (AIPM) contractors should be required to accelerate the development of physical prototypes of their design concepts.

Recommendation. The electronics and electrical systems technical team should assure that there is effective communication between the automotive power module (AIPM) and automotive electric motor drive (AEMD) developers and the organizations engaged in material and component development.

STRUCTURAL MATERIALS

From the outset of the PNGV program the reduction of vehicle mass was recognized as one of the key strategic approaches in meeting PNGV Goal 3. To achieve Goal 3's 80-mpg target, systems analyses showed that a 40 percent vehicle weight reduction was necessary, together with additional measures such as 40 to 45 percent power-train thermal efficiency, 70 percent efficient regenerative braking, improved drive-line efficiency, and reduced aerodynamic drag. This must be achieved while maintaining the baseline vehicle performance, size, utility, and affordable cost of ownership.

Substantial vehicle-weight reduction targets for various subsystems have been set (Table 2-8). Achieving these targets will result in an overall reduction in curb weight of the baseline vehicle of about 2,000 lb (40 percent reduction).

Materials Selection, Design, and Manufacturing

In the search for lightweight materials that would allow the targeted large weight reductions (50 percent for the body-in-white, for example), PNGV placed heavy emphasis on materials whose density is substantially less than the steels used in the baseline vehicle (NRC, 1998, 1999). This is particularly evident in comparing the attributes of the PNGV 2000 concept vehicles as shown in Table 2-9. It is interesting that, although one of the concept vehicles made the 80-mpg target and the others came close, none of the vehicles made the weight target, in spite of the heavy use of low-density materials. Also, it appears likely that there will be great difficulty in meeting the PNGV affordability target by 2004.

The high cost of low-density materials and associated manufacturing costs is one of the primary reasons that it will be difficult to meet the PNGV affordability target, as can be seen in Table 2-10. The estimates in relative cost given in

TABLE 2-8 Weight-Reduction Targets for the Goal 3 Vehicle

Subsystem	Current Vehicle (lb)	PNGV Vehicle Target (lb)	Mass Reduction (%)
Body	1,134	566	50
BIW[a]	590	—	—
Chassis	1,101	550	50
Power train	868	781	10
Fuel/other	137	63	55
Curb weight	3,240	1,960	40

[a]Body-in-white (BIW) includes all the structural components of the body, the roof panel, and the subframes, but not the closure panels.

SOURCE: Adapted from Stuef, 1997.

TABLE 2-9 Selected Attributes of PNGV 2000 Concept Vehicles

Attributes	PNGV Targets	DaimlerChrysler Dodge ESX3	GM Precept Hybrid	Ford Prodigy
Cost/ affordability	Equivalent to current vehicles	$7,500 price premium	N/A	Not affordable
Estimated fuel economy[a]	Up to 80 mpg	72 mpg	80 mpg	70 mpg
Curb weight	898 kg (1,980 lb)	1,021 kg (2,250 lb)	1,176 kg (2,590 lb)	1,083 kg (2,385 lb)
Body structure	N/A	LIMBT[b] on aluminum space frame	Aluminum space frame & panels, and CFRP[c] sheet	Aluminum unibody

[a]Fuel economy is in miles per equivalent gallon of gasoline.
[b]LIMBT = lightweight injection-molded glass-fiber-reinforced polymer (GFRP) body technology.
[c]CFRP = carbon-fiber-reinforced polymer.

TABLE 2-10 Weight Savings for Lightweight Materials

Lightweight Material	Material Replaced	Mass Reduction (%)	Relative Cost (per part)[a]
High strength steel	Mild steel	10-24[b]	1
Aluminum	Steel, cast iron	40-60	1.3-2
Magnesium	Steel or cast iron	60-75	1.5-2.5
Magnesium	Aluminum	25-35	1-1.5
Glass FRP[c]	Mild steel	25-35	1-1.5
Carbon FRP[c]	Mild steel	50-65	2-10+
Aluminum MMC[d]	Steel or cast iron	50-65	1.5-3
Titanium	Alloy steel	40-55	1.5-10+
Stainless steel	Mild steel	25-40	1.2-1.7

[a]Includes both materials and manufacturing costs; the lower bound of unity is a future projection.
[b]The lower figure is taken from Powers (2000) and the upper bound from NRC (2000).
[c]FRP = fiber-reinforced polymer.
[d]MMC = metal matrix composite.
SOURCE: W.F. Powers, 2000.

Table 2-10 are supported by an independent study that concluded that use of an aluminum body-in-white (BIW) and closure panels leads to an incremental cost of $1,400 (Schultz, 1999) over the baseline steel BIW and closure panels, depending on several variables.

Another approach to reducing vehicle weight has been taken by the steel industry (Jeannes and van Schaik, 2000; NRC, 2000). The Ultralight Steel Auto Body (ULSAB) project explores the weight reductions that could be achieved through (1) greater use of higher-strength steels than in the baseline vehicle and other approaches, such as steel and plastic sandwich structures; (2) finite element modeling; and (3) innovative manufacturing processes, such as laser-welded tailored blanks and hydro-formed tube structures and roof panels (ULSAB, 1999).

All these processes were integrated into a unified approach to reduce the average steel sheet thickness and thereby save weight. At the end of the study the ULSAB BIW weighed only 447 lb, a 24 percent reduction over the PNGV baseline vehicle. The study also concluded that the ULSAB BIW would be $154 less costly than the baseline BIW.

The American Iron and Steel Institute (AISI) has embarked on a follow-on study to ULSAB, involving an Advanced Vehicle Concept (ULSAB-AVC), which will result in complete design concepts for an ultra-light steel-intensive car that meets projected 2004 vehicle and crash requirements (NRC, 2000). The ULSAB-AVC vehicle will be a PNGV-class vehicle (i.e., a 5-passenger, 4-door sedan) targeted to have an overall length of 187 in (4,750 mm) and a total weight of 2,275 lb when powered by a gasoline-fueled internal combustion engine (ICE). In addition, AISI has projected what the safety standards might be in 2004 and from this projection determined that an additional 55 lb of steel structures will be required to meet this requirement. Considerable progress has been made during the past year. Concepts for the BIW, closures, suspension system, and engine system have been designed and modeled. The design phase should be completed in 2001.

Thus far it appears that the ULSAB-AVC concept is not targeted to include a hybrid diesel and electric power train, which will make it difficult to make a direct comparison between the efficient steel concept and aluminum-intensive vehicle approach. Because steel is the least costly solution, it is prudent for the PNGV team to look for ways to evaluate the unique designs and processes that may have application to lighter metals as well, and to look for ways to increase the weight-reduction potential of this approach.

Materials Roadmap

The PNGV materials team has developed a materials roadmap that identifies lightweight material alternatives for the major subsystems of the vehicle. The roadmap is a multi-year plan and has been summarized in previous committee reports (NRC, 1999, 2000). The process used in selecting candidate materials for the roadmap involved analyzing several materials and processes for each major component (e.g., body structures, body panels, suspension springs, engine components). The span of alternative materials and processes considered was quite broad, ranging from next-generation materials such as plastics, aluminum, and magnesium to aerospace materials, such as titanium alloys and carbon-fiber-reinforced polymers (CFRPs). Generally speaking, the very-long-range alternatives, such as titanium alloy and CFRP components, are very costly. These long-range candidates, however, remain in the roadmap because there are R&D programs aimed at reducing their feedstock costs. With respect to point 3 of the committee's statement of task (see Chapter 1), the committee believes that, based on its own knowledge of worldwide materials developments and extensive benchmarking conducted by the PNGV materials technical team in preparing the materials roadmap, it is very unlikely that the PNGV effort is likely to be blindsided by a new materials technology.

While weight savings have been a prime consideration in choosing R&D work, programs aimed at reducing the cost of low-density materials are a major part of the project portfolio. All areas of cost are being considered, such as improved designs, reducing feedstock costs through alternative processing, reducing part fabrication costs through alternative processes and tooling, and recycling. The synergy that results from taking this holistic view is depicted in Figure 2-1.

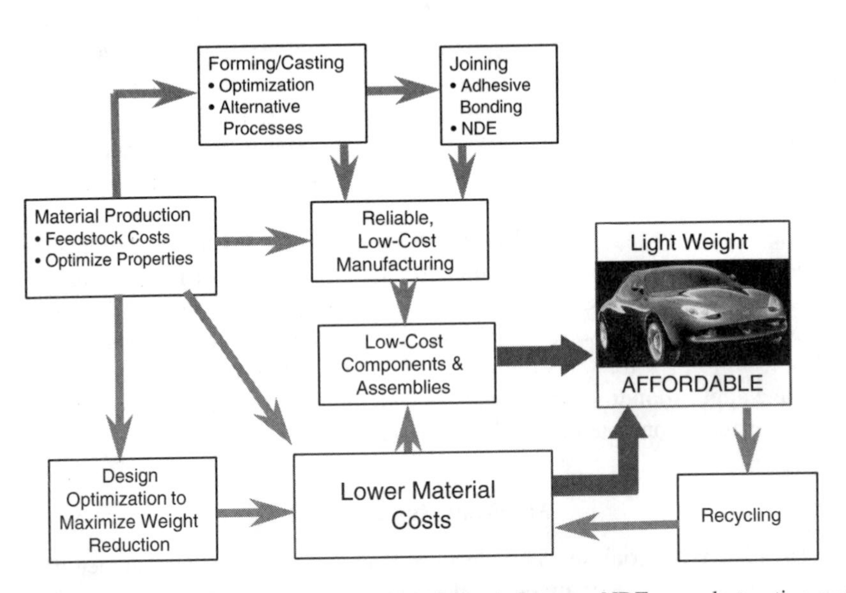

FIGURE 2-1 Lightweight materials: affordability influences. NDE = nondestructive evaluation.
SOURCE: Mehta, 2000.

Over 30 materials projects have been established to attack the technical challenges identified in the materials roadmap. Some projects have been completed; others are in progress, and some have just been added. The committee believes that these projects have the potential to remove the barriers that make it difficult to implement alternative materials in today's vehicles.

Program Status

Completed Projects

In a project on Design and Product Optimization for Cast Light Metals, a user-friendly design guide for chassis components was developed, based on solidification simulation models that accurately predict cast microstructure and the resulting mechanical properties. When applied to a cast aluminum control arm, these techniques led to a 26 percent weight savings, reduced cycle time, and cost savings of $1.90 per part.

A project on Stamping Press Optimization of Aluminum Sheet demonstrated the viability of variable blankholder force in improving formability and reducing wrinkling and splitting in simulated and actual auto fenders. This leads to reduced scrap and a reduced need for die fine tuning through welding and grinding, and enables parts reduction and consolidation. A project on Warm Forming of Aluminum Sheet demonstrated the ability to form a complex aluminum door inner panel not possible using traditional stamping. This was accomplished by heating the dies to ~250°C. This results in reduced scrap due to wrinkling and tearing and eliminates the need for a re-draw stage required in steel.

The Low Cost Aluminum Sheet project to develop 5000-series continuous cast sheet for body structures was successfully completed last year. Belt-cast thin sheet, produced with several process or alloy variations, was used successfully to form several large and challenging prototype parts. This process for producing lower-cost aluminum is being considered by major producers of aluminum sheet.

In the area of polymer composites a High Volume Liquid Composite Molding project (Automotive Composite Consortium Focal Project 2) has demonstrated the ability to produce high-volume preforms (the P-4 process), the ability to mold large structural components, and the viability of structural adhesive bonding. A polymer composite truck pickup box was developed by liquid composite molding, resulting in a weight savings of 27 percent for the complete assembly (see Figure 2-2). A cost model was developed that shows that the composite pickup box is lower in cost relative to steel for target volumes in the range of 50,000 to 75,000 units, as shown in Figure 2-3.

The first application of this composite technology is the pickup box for the 2001 Chevrolet Silverado, which was named one of the Ten Best Innovations of the Year by *Popular Science* magazine. The box is 50 lb lighter and the tailgate is 15 lb lighter than their steel counterparts. The impact and corrosion resistance of

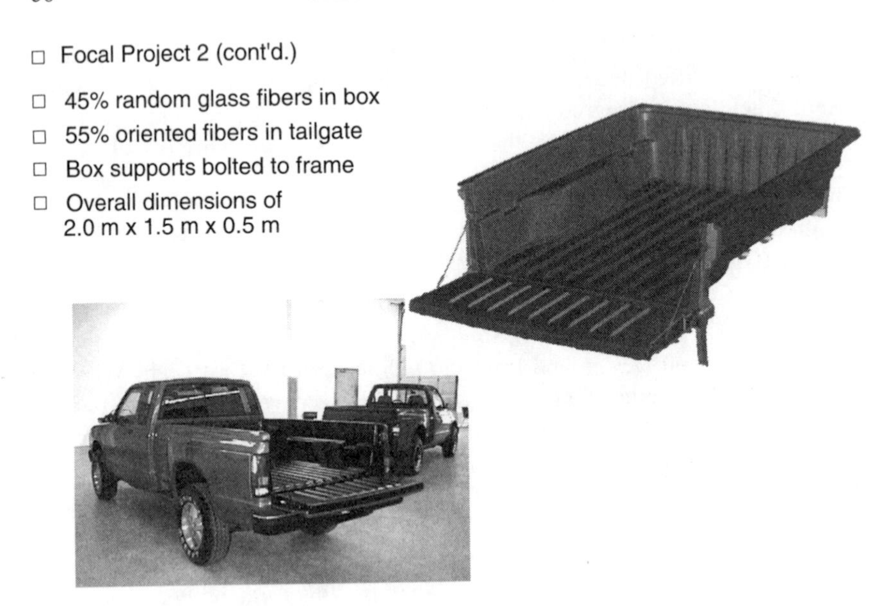

FIGURE 2-2 Polymer composite pickup box.
SOURCE: Mehta, 2000.

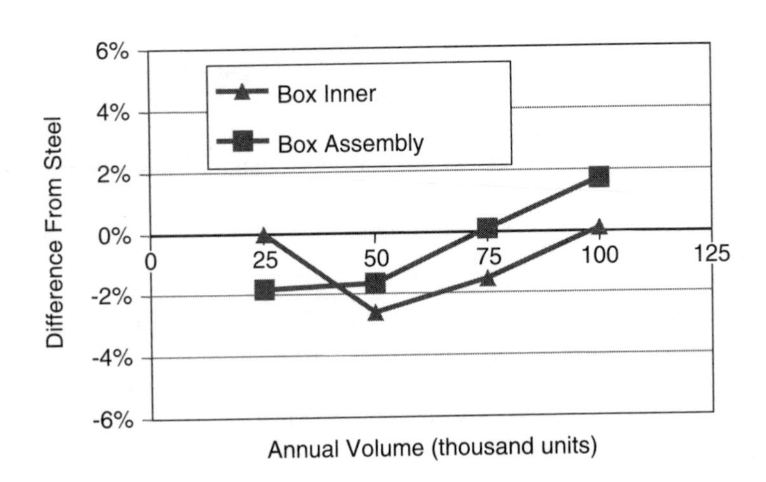

FIGURE 2-3 Cost of polymer composite pickup box relative to the cost of steel pickup box.
SOURCE: Mehta, 2000.

the box is far superior to that of steel truck beds. Also, Ford has used sheet molding technology to produce an all-composite, one-piece SUV truck bed for the Ford Explorer Sport Trac. The steel counterpart was 30 percent heavier and composed of 40 separate parts.

Another approach to composite body technology is being pursued by DaimlerChrysler in its proprietary PNGV program (NRC, 1999, 2000). Its process produces large, one-piece parts by injection molding of glass-fiber-reinforced polymer (GFRP) resins. The integrated moldings, which are 25 to 35 percent lighter than mild steel, are attached to an aluminum space frame, resulting in a BIW 50 percent lighter than steel (NRC, 2000). As indicated in Table 2-9, this technology was used in the DaimlerChrysler ESX3 concept vehicle, which achieved the lowest curb weight of the three concept vehicles developed by the USCAR partners (see Table 2-9).

Projects Under Way

Focal Project III (FP III) supported by the Advanced Composites Consortium (ACC) involves the development of BIW structure in GFRP composites. This entails constructing the body from approximately 11 large liquid molded parts, as shown in Figure 2-4. Thus far, computer-aided design (CAD) and coarse finite element analysis (FEA) models have been generated for initial concepts

Key Accomplishments of FP III:

- Generated CAD and coarse FEA models for initial concepts.
- Evaluated initial concepts based on performance, processing, and assembly.
- Completed bending and torsional stiffness optimization of a series of concepts of body-in-white structures.
- Generated detailed finite element models of chosen concept for analysis of all load cases - including stiffness, strength, durability, and modal performance.
- Identified key processing and material options that require further development in order to manufacture the FP III body-in-white structure.

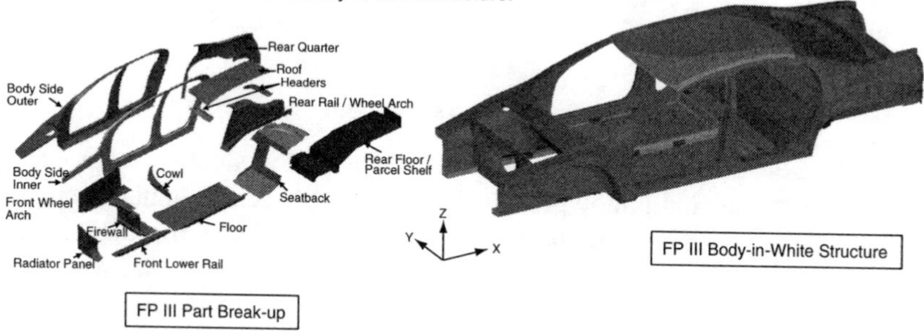

FIGURE 2-4 The ACC Focal Project III body-in-white structure.
SOURCE: Mehta, 2000.

and have been evaluated for performance, processing, and assembly. Bending and torsional stiffness of the BIW structure have been optimized. Detailed finite element models of chosen concepts have been generated for analysis of all load cases, including stiffness, strength, durability, and modal performance. Key processing and material options that require further development in order to manufacture the FP III structure have been identified.

In the past, work by the ACC showed that it was difficult to achieve good frontal crash response from polymer composite front-end structures. These early problems have been overcome. The ACC's Focal Project 1 demonstrated that a composite structure can be designed to meet Federal Motor Vehicle Safety Standard (FMVSS) requirements. In this case the front end of an Escort was built and successfully tested in a frontal crash experiment.

Although the ACC projects are aimed at glass FRP, where weight savings are in the range of 25 to 35 percent, the technology of liquid molding of composites can be adapted to carbon FRP, where weight savings approach 65 percent. Projects are in place with suppliers and ORNL to develop low-cost carbon fibers through lower-cost precursors and by microwave processing. Potential exists to reduce fiber cost by 20 percent.

Aluminum metal matrix composite (Al-MMC) applications in chassis and power-train subsystems could account for 30- to 50-lb weight savings. The major hurdles to developing applications of this material are feedstock costs (Table 2-10) and the development of a reliable process for making the composites. A low-cost powder metal process and a casting process are under development. In the casting process a new and improved batch technique for mixing silicon carbide in the aluminum melt has been developed, and the performance of a lower-cost (silicon carbide) reinforcement has been demonstrated. Equipment that can prepare a 600-kg sample batch of composite material was expected to be operating in January 2001.

In the area of recycling, a project on Aluminum Alloy Scrap Sorting has been initiated to demonstrate the commercial practicality of sorting shredded scrap, using color and laser spectroscopy. This process is capable of separating cast from wrought materials and distinguishing among wrought alloys.

Program Assessment

As the PNGV program moves toward 2004 and the requirement of producing affordable production-prototype vehicles, the PNGV team should attempt to balance the opposing requirements of weight reduction and affordability. The arguments presented in the "Materials Selection, Design, and Manufacturing" section above lend credence to this position. The PNGV team should carefully follow the progress on the new ULSAB-AVC project. This modeling study appears to be on track to demonstrating that a 2,275-lb steel-intensive vehicle, operating on a gasoline-powered ICE engine, can be attained. It may be possible to add aluminum and/or magnesium components (hybridize) and/or polymer composites and achieve the vehicle weight target at a much lower cost than with the aluminum-intensive vehicle approach. These design and manufacturing techniques could be used on aluminum-intensive vehicles as well as hybrid steel-aluminum structures.

The challenge in being able to adopt low-density materials on a broad scale is the development of new, low-cost processing methods, including feedstock materials. The committee is satisfied that the PNGV materials technical team has developed a portfolio of R&D programs that is directed toward producing lower-cost, low-density materials. Having said this, the committee finds it difficult to track progress on a given project on a year-to-year basis. Each year the PNGV materials technical team selects a different group of projects on which to report. For example, last year magnesium and titanium projects were discussed, but this year they are barely mentioned. An alternate reporting approach is needed.

Recommendations

Recommendation. Because affordability is a key requirement of the 2004 production-prototype vehicles, the committee believes that more attention should be paid to the design and manufacturing techniques being worked on by the American Iron and Steel Institute in the Ultralight Steel Auto Body Advanced Vehicle Concept project. These techniques should be applied to aluminum-intensive vehicles, as well as hybrid-material body construction. More broadly, the committee urges a systematic, critical examination of the prospects to achieve cost goals for all key vehicle subsystems and components.

Recommendation. Given the difficulty of tracking progress on more than 30 materials R&D projects, PNGV should devise a new reporting approach that measures progress against objectives.

VEHICLE SAFETY

The economic feasibility of PNGV-class vehicles will depend on meeting or exceeding consumer expectations in vehicle function, comfort, and safety. The Goal 3 safety requirement in the 1995 PNGV Program Plan was to meet all Federal Motor Vehicle Safety Standards (PNGV, 1995). Since 1995 consumer choice has shifted to heavier vehicles that also are perceived to provide increased safety. The PNGV vehicle is designed to replace a heavier passenger car of the same size in a fleet that is dominated by even heavier cars and light trucks. This environment may be an impediment to selling these vehicles. Also, since 1995 family vehicles have introduced a profusion of safety features beyond those required by the government, some of which may have to be incorporated into these new vehicles for them to be competitive. In addition to these considerations, the new technologies, introduced by PNGV-type cars, will create new failure modes and safety concerns that will have to be considered.

Status and Progress

The PNGV concept cars were designed to meet all existing standards with the exception of FMVSS 111, which requires outside rear-view mirrors (CFR, 2000). To reduce aerodynamic drag on these vehicles, outside mirrors were replaced with an indirect viewing system that uses digital cameras and a flat screen display inside the vehicle. These designs offer opportunities to increase the amount of information available to the driver. However, it is not known how well a flat screen display will replace the external mirror presentation to the driver under all conditions. Research by the PNGV is planned during 2001 to assist the National Highway Traffic Safety Administration (NHTSA) in evaluating human factors associated with alternatives to exterior mirrors. The results will provide NHTSA research information to assist in a possible upgrade to FMVSS 111.

The 1995 PNGV Program Plan calls for NHTSA involvement to "help assure that any vehicles offered for sale possess structural integrity, include occupant protection systems, and do not compromise safety levels" (PNGV, 1995). In December 1999 the Safety Working Group was established to involve the PNGV industry partners and NHTSA more actively in safety issues. The working group has identified and prioritized five safety research needs by PNGV-class vehicles.

The working group has recognized the need to address safety concerns that extend beyond the goal of meeting the minimum FMVSS. These issues involve the safety of PNGV-related designs that may not be addressed by present standards. For example, in general, customers associate lighter vehicles with decreased safety. This may impede the sales of PNGV-type vehicles. It has been well documented that smaller, lighter vehicles provide less occupant crash protection (DOT, 1997). The PNGV-type vehicle will use lighter materials and new construction techniques to reduce weight but maintain the vehicle size. Consequently, the effect on occupant safety may be mitigated due to increased crush space available. This important issue is being addressed by analyzing field accident data and is funded by the industry partners. The study will examine the relationship between safety and vehicle size, independent of vehicle weight. The absence of PNGV-type vehicles in the database makes it more difficult to separate the effects of vehicle size from vehicle weight. A second study of vehicle size and weight that uses computer modeling of vehicle structures in crashes has been defined but not funded. In its sixth report the committee recognized this research need and recommended that it be accelerated (NRC, 2000).

Other as yet unfunded safety research projects identified by the Safety Working Group include (1) mathematical models for joints in integrated materials structures; (2) HEV power trains (power electronics, energy storage); and (3) alternative fuels.

Changes in Environment

The PNGV safety goals developed in 1995 apparently did not anticipate the changes that have subsequently occurred in consumer attitudes toward vehicle weight and safety. Most consumers (and some physicists) equate increases in vehicle weight with increases in safety. In today's market a lightweight vehicle like the PNGV-type vehicle may be unacceptable to safety-conscious consumers. However, the overall safety of a vehicle is not governed totally by the vehicle's weight. Improved crash-avoidance features, higher levels of crash protection, and increased vehicle size can offset the weight disadvantage. The marketability of the PNGV-type vehicle will depend on demonstrating to consumers that they are not trading safety for fuel economy. To date, the PNGV project has not adequately addressed this issue. In addition, since 1995 a large number of new safety features beyond those required by federal standards have been introduced. These features were supported by a competitive market rather than government mandates. Examples of market-driven safety features on today's family vehicles include anti-lock brakes; automatic traction control; automatic stability control; low-tire-pressure warning; and many new occupant restraint devices. By the time PNGV-type vehicles enter the marketplace, the level of vehicle safety technology will be quite different from the 1994 family sedans that were the baseline for PNGV function and safety, and this will have to be reflected in the ultimate designs.

As noted above, consumer choice has shifted during recent years to heavier vehicles that are also perceived to provide increased safety, and marketing for PNGV-type vehicles will have to deal with this perception. In addition, the introduction of new technology has occasionally created new safety concerns, such as unintended injuries and fatalities from air bag deployments in unusual circumstances. These concerns include not only the crashworthiness of new designs and materials but also new safety issues in flammability, fire and explosion, toxicity, sources of heat and chemical burns, and electrical shock. Defining the safety issues associated with the new technologies for PNGV-type vehicles will be a priority of the recently established Safety Working Group. The committee believes that funding for this work will have to be increased to adequately address these issues. Also, research support from the government appears to be nonexistent in this area, which should be corrected.

Recommendations

Recommendation. Because of the unresolved issues related to safety, the safety research in support of PNGV should receive a higher priority. Critical projects identified by the Safety Working Group should be accelerated and funded appropriately. More effective research support should be provided by the U.S. Department of Transportation.

Recommendation. Any initial low-volume production fleet vehicles should be monitored for early signs of any safety issues.

FUEL ISSUES

All the power system options under evaluation in the PNGV program appear to require fuel modifications that would have significant impacts on the petroleum industry, as pointed out previously (NRC, 2000). Mindful that fuels compatible with efficient engine technologies must be widely available to produce meaningful reductions in fuel use, the PNGV has continued to investigate the interactions of power systems and fuels. There have been important additional testing programs involving auto and petroleum companies, including the Ad Hoc Diesel Fuel Research Program, the Coordinating Research Council (CRC) Advanced Vehicle, Fuel, Lubricant (AVFL) Committee, and the Advanced Petroleum-Based Fuels (APBF) Program, as well as efforts through the CARB Fuel Cell Program, and EUCAR/USCAR Cooperative Fuels Research (see Chapter 2 section, "Internal Combustion Reciprocating Engines"). The primary power plant options under consideration in the PNGV are the CIDI engine in the HEV configuration and fuel cells; the fuel implications of each of these are discussed below.

Fuels for CIDI Engines

The CIDI engine continues to be potentially attractive because of its high efficiency, but emissions of PM and NO_x are of concern, particularly in view of the EPA Tier 2 standards. PNGV is pursuing options for reducing engine-out emissions, as well as after-treatment methods. Fuel composition influences both of these approaches.

Sulfur content is probably the most important fuel parameter, affecting both engine-out emissions (primarily particulates) and the performance of after-treatment systems to reduce either NO_x or particulates. In the past year priority in the program has been on identifying systems with the potential to meet the Tier 2 standards and, while available data indicate that any sulfur is detrimental, the PNGV has not determined just how low the sulfur content must be to ensure long-term operation. Results show that reducing sulfur improves the effectiveness of SCR systems for NO_x control. Work on sulfur traps has continued and indicates that such traps might make it feasible to use sulfur-sensitive after-treatment with sulfur-containing fuels. Upstream trapping of sulfur made it possible to achieve 90 percent reduction of NO_x, using a NO_x adsorber, which is very sensitive to sulfur. However, available data from Toyota indicate that the trap regeneration temperature will be on the order of 650°C for periods of 10 minutes or more, and even then satisfactory regeneration was obtained only with a feed sulfur content of 8 ppm (Toyota, 2000). Much more work on this concept is required to provide relief from the feed sulfur restriction. Sulfur traps would require periodic maintenance with a frequency depending on the fuel sulfur level, and this would raise the issue of ensuring consumer participation in such maintenance.

Lower-sulfur commercial fuels will be available by June 1, 2006, because of the EPA (Federal Register, 2001) final rule requiring refiners to produce diesel fuel with a maximum sulfur content of 15 ppm. Based on answers to committee questions, the PNGV expects that, to ensure 15 ppm maximum sulfur, the average sulfur content will be 10 ppm (PNGV, 2001). A recent study (DOE, 2000a) indicates that less than 10 ppm sulfur may be required to achieve "minimum acceptable effectiveness" of CIDI engine emission control devices. Nevertheless, in discussions with the committee, PNGV members expressed the view that an average sulfur content of 10 ppm and a maximum level of 15 ppm would be marginally acceptable. If current efforts fail to develop systems that can tolerate the level of sulfur mandated by EPA, a major program re-assessment will be required.

As addressed in the section "Internal Combustion Reciprocating Engines" in Chapter 2, fuel composition can impact the production of regulated emissions, most notably PM and NO_x. It has been concluded that, even under the most favorable conditions, the reduction of in-cylinder emissions is not sufficient to preclude extensive exhaust-gas after-treatment. In addition to the assessment of fuel composition impacts on regulated emissions, PNGV also investigated the impact of fuel composition on potential health-impacting toxic chemicals. One program completed in 2000-2001 was directed toward chemical characterization of engine-out diesel emissions, including compounds that might have toxicological effects,[4] using advanced fuels. This study showed that use of a Fischer Tropsch "diesel fuel" low in aromatics, or a mixture of an oxygenate and a low-sulfur petroleum fuel, significantly reduced the emissions of hydrocarbons, particulate matter, polyaromatic hydrocarbons (PAHs), and aldehydes, when compared with emissions using conventional diesel fuel. In addition, these fuels produced the lowest overall "toxic" gas and PAH exhaust emissions. Thus, avail-

[4]Emissions of 4 potentially toxic gaseous emissions, 11 gaseous PAH compounds, and 17 particulate-soluble organic-phase PAH compounds were measured.

able data indicate that gross changes in fuel composition (i.e., reducing aromatics content or adding oxygenates) can affect engine-out emissions in significant ways.

The PNGV told the committee that a fuel cetane number of at least 45 will probably be required, and this will restrict the aromatics content to some extent. The committee notes that reducing sulfur to 15 ppm could lead to a reduction in aromatics content and an increase in cetane number since the hydro-treating process used to reduce sulfur will also directionally saturate aromatics. No definitive data on the extent of this effect are available to the committee.

Another program determined the contribution of lubricating oil to the emissions of PM from an advanced diesel engine. Lubricants studied included synthetic and mineral-oil-based materials. CARB diesel and a mixture of an oxygenate and low-sulfur petroleum were used as fuels. These lubricants had a smaller effect on particulate emission than the fuels studied. However the lubricant contribution could become important at very low exhaust emission levels.

The supply of aqueous urea would quickly become an important issue if urea SCR were to become the preferred NO_x-control strategy. While this is not a fuels issue, it is mentioned here since fueling stations would be one potential means of providing this urea. Pursuing this possibility would require the cooperation of the petroleum industry.

The committee believes that the most important fuel issue is the sulfur content required to meet long-term emission certification. If the preferred power systems were to require sulfur levels lower than those mandated by EPA or other compositional changes, additional government action would be needed to coordinate the efforts of the auto and petroleum industries to ensure timely availability of commercial fuel. The committee believes that a very important issue is the maximum sulfur level required to meet the long-term emission certification.

Fuels for Fuel Cells

While some work has been initiated on the direct methanol fuel cell, the PNGV has focused primarily on the hydrogen fuel cell with a gasoline fuel processor to provide the hydrogen. The program also includes some work on a flexible-fuel processor as well as distributed hydrogen generation at service stations, combined with onboard storage of hydrogen. This program is developing technologies that could be applied to different fuels, but in applications the fuel processor would be optimized for one fuel only.

The reforming catalyst that converts gasoline to hydrogen is sensitive to sulfur. While progress has been made to increase the sulfur tolerance of this catalyst, it is not clear what sulfur level will be acceptable. In addition, reforming is more difficult with more aromatic gasolines. The systems under development might require a reduction of sulfur and aromatics content; in fact, the required aromatics content might be low enough to require a different petroleum fuel, such as naphtha, as the committee pointed out last year in its sixth report (NRC, 2000). A presentation on the California Fuel Cell Partnership indicated to the committee a lack of optimism for the ultimate success of the gasoline (or modified gasoline) reformer (Wallace, 2001).

Because of the issues with gasoline and the fact that methanol is much easier to reform to hydrogen, reforming of methanol or other oxygenates continues to be in the program. A multiyear program (DOE, 2000b) to address these issues and to explore the use of reformulated diesel, methanol, ethanol, and Fischer-Tropsch liquids in fuel-flexible fuel processors is under way.

While the committee feels that additional efforts are needed to identify fuel and reformer combinations that are feasible, it is not optimistic about the future for fuel-flexible processors, in view of the added complexity introduced by the requirement for multiple fuels. The program is comprehensive in scope. As results are obtained, it will be important to narrow the scope quickly, using decision points that are included in the program plan. The study should include an update of the cost and availability outlook for methanol and other oxygenates based on an updated analysis of the long-term supply and cost of natural gas and from Fisher-Tropsch fuels. As pointed out by the committee last year (NRC, 2000), the capital cost for hydrogen generation at large, central facilities followed by wide-scale distribution is roughly $3,500 to $6,700 per vehicle.[5] Last year the committee recommended consideration of distributed hydrogen generation at service stations, and DOE has included this in its program mentioned above. The committee notes that reforming of gasoline or methanol is not a clear winner to reduce carbon dioxide emissions and increase overall efficiency of fuel use. As options for off-vehicle hydrogen generation are evaluated, the effects of each approach on life-cycle efficiency and carbon dioxide emissions should be kept in focus. The DOE program also includes work on chemical storage systems, such as metal hydrides, which have the potential to reduce onboard storage volume or storage pressure, or, with additional storage and associated weight, increase the vehicle's driving range.

In view of the difficulties with supplying hydrogen for the hydrogen fuel cell, it is significant that DaimlerChrysler announced in 2000 its first test vehicle, a go-cart equipped with a 3-kW direct methanol fuel cell (with no reformer) (Muller, 2000). The vehicle is reported to have a top speed of 22 mph, and the company has already built a 6-kW, 60-V version. This represents an important step along the way toward development of fuel-cell-powered vehicles that do not require hydrogen. The company estimates that direct-methanol fuel cell vehicles could be available in about 10 years.

---

[5]In comparison, the onboard gasoline processor would cost $4,350 per vehicle based on PNGV targets (a 50-kW system at a cost of $300/kW), 29 percent of which is for the fuel processor, and would eliminate the need for a hydrogen distribution system.

As pointed out last year, methanol is not widely available at the present time, but it could be distributed through existing service stations with modifications to the supply and distribution system, for example, to maintain product segregation, manage water, and replace any materials not compatible with methanol. Issues related to its corrosive properties and potential public health effects would require further investigation, and time would be required for fuel suppliers to make the facility modifications necessary to supply an additional fuel. In addition, new plants to manufacture methanol could be required depending on the extent of commercial use of this option, and methanol cost would be affected by natural gas cost. On the other hand, because methanol is used to make methyl tertiary-butyl ether (MTBE), if the use of MTBE in gasoline decreases, more methanol could become available without new plants.

It is clear from the foregoing discussion that the hydrogen required by fuel cells could be generated from gasoline, another petroleum liquid fuel, methanol, or natural gas. Some options call for onboard generation, and others, for generation in centralized facilities. Given the program goals of increasing efficiency and reducing fuel consumption, it will be important to make all-inclusive, "well to wheels" analyses (Weiss et al., 2000) to account for all factors associated with providing hydrogen.

The present DOE Hydrogen Program has a total of 77 projects that deal with hydrogen production, storage, and use for both stationary source applications and vehicular use (DOE, 2001).

Recommendations

Recommendation. High priority should be given to determining what fuel sulfur level will permit the preferred compression-ignition direct-injection (CIDI) engine and its after-treatment system to meet all regulatory and warranty requirements. An enhanced cooperative effort between the auto and petroleum industries should be undertaken to ensure that the fuels needed commercially will be available on a timely basis.

Recommendation. Given the breadth of the multiyear Fuels for Fuel Cells R&D Program, the go/no-go decision points should be closely followed to facilitate identification and timely development of the preferred options.

3

Vehicle Engineering Developments

SYSTEMS ANALYSIS

Systems models integrate models of individual vehicle and power-train components to predict component and overall vehicle system performance. They are used to examine the effect of varying individual component design and performance characteristics on overall system performance. Thus, they are used to develop optimum system configurations and the component performance specifications for such configurations for various types of vehicles over various driving patterns. They can also provide information on system cost and reliability, using hypothetical or actual future components, to perform trade-off studies.

The PNGV Systems Analysis Team has developed the PSAT (PNGV Systems Analysis Tool) model over the past several years for these purposes. This tool is now well developed and is being used to further the objectives of the PNGV program. PSAT is a driver-driven or forward model in which the performance of the vehicle and of its components is determined based on vehicle driver inputs. In 2000, DOE assumed responsibility for the ongoing development of the PSAT model in parallel with another earlier vehicle systems model, Advisor, developed by the National Renewable Energy Laboratory (NREL). Advisor is a backward, or drive-cycle-driven, model in which the power-train performance is calculated from the torque and speed requirements needed to drive the vehicle through the specified drive cycle.

In its last report, the committee recommended greater emphasis on validating PSAT component and overall vehicle predictions, developing emissions modeling capabilities, developing a generic system and subsystem cost model, and increasing fuel cell component and system modeling (NRC, 2000). During 2000-

2001, substantial emphasis was placed on validating the PSAT model, using data from the Toyota Prius and Honda Insight HEVs and the Ford P2000 fuel cell vehicle. Vehicle testing and simulation evaluations are still in progress, and the results to date are encouraging.

Major efforts have been initiated on modeling emissions from internal combustion engines and their exhaus-gas after-treatment systems (catalysts and particulate traps). While this is a challenging technical task, the initial progress is encouraging, and approaches have been developed that should provide useful estimates of vehicle emissions to assess compliance with future emissions standards. It is apparent that the challenge of modeling the effects of engine acceleration and deceleration transients on vehicle emissions with HEV systems is less severe than with stand-alone engine transient modeling because the HEV system reduces the impact of these vehicle drive transients on the engine. However, the HEV internal combustion engine undergoes many start-ups and shut-downs during normal driving. Modeling the emissions produced by these events is especially challenging. The committee is encouraged by this progress in emissions modeling and recommends continuing emphasis on this topic. One important use for a vehicle emissions model is to quantify the trade-off between efficiency and emissions for the various propulsion system options, as discussed in Chapter 5.

Fuel cell HEV system simulation studies of a large sport utility vehicle (SUV) are under way to examine the trade-off in relative sizing of the battery pack and fuel cell. This indicates that a useable fuel cell HEV system model is now available. An especially important fuel cell system modeling area that was not part of the fuel cell modeling review is the liquid fuel reformer system. An effective gasoline-to-hydrogen reformer is a critical component in the most practical shorter-term fuel cell system because it avoids the challenges of developing a hydrogen production and distribution system and the need to store hydrogen on the vehicle. The committee encourages increased effort in modeling fuel cell system-component and overall system performance and especially in the fuel-reformer technology area. Because fuel cell technology is developing rapidly, well-validated system models are important tools for extrapolating from the performance of current prototype system data to likely future system performance.

Progress in cost modeling was not as encouraging as in the areas described above. However, a Cost Analysis Task Group has recently been formed. The committee urges that a framework be developed for using the systems model to assist in cost estimation studies for the internal combustion engine (ICE) and fuel cell HEV systems, and the effort on detailed modeling required to implement effective cost models should be intensified.

HYBRID PRODUCTION VEHICLES

Since the last committee report (NRC, 2000), both Toyota and Honda have introduced HEVs into the U.S. market. An overview of these two vehicles illustrates the wide range of design philosophies and component configurations that are classified as HEVs. It also points out the differences between the Toyota and Honda hybrids and how they compare to developments being pursued by the USCAR partners.

In the simplest terms, an HEV power train can be defined as a power propulsion system in which both an engine and an electric motor work together to propel the vehicle. Hybridization allows three distinct design principles to be incorporated into the vehicle: engine downsizing (by using electric power assist), battery-only electric driving, and regenerative braking. Additional approaches to improve efficiency, such as engine stop/start at idle and operating along the best efficiency versus power curve, may be incorporated into hybrid power trains, but they are not exclusive to them.

The Honda HEV (the Insight) and the Toyota HEV (the Prius) are smaller vehicles than the PNGV development vehicle (the family sedan). The Prius is advertised as a five-passenger sedan, albeit small, while the Honda Insight is a two-passenger sedan. The Prius has four doors and weighs approximately 2,700 lb, while the Insight is a two-door vehicle that has very limited interior space and weighs approximately 2,125 lb. Both use relatively conventional body and structural design and materials. Although exact numbers are held in confidence, it is generally accepted that both vehicles are being sold at a loss to the respective parent companies.

An examination of the power-train configurations of the two vehicles helps to illustrate the wide range of power-train configurations that are possible under the HEV classification, as well as the different optimization schemes that can be employed for improving fuel economy (see Figures 3-1 and 3-2). Both vehicles are parallel hybrid configurations; that is, in principle the vehicles could be driven either by the engine or the electric motor or both.

Gas mileage tests of the two vehicles have recently been performed and reported by Argonne National Laboratory (ANL) (Duoba and Ng, 2001; Duoba et

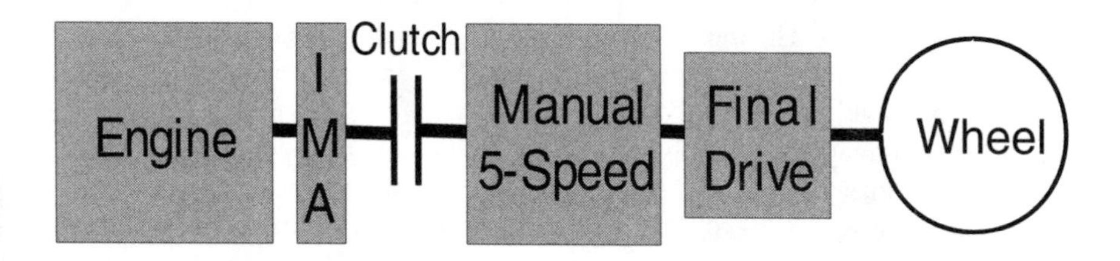

FIGURE 3-1 Honda Insight power-train configuration. The components shown are the energy converters, not storage devices. The batteries are not shown. IMA = integrated motor assist.
SOURCE: Duoba et al., 2001.

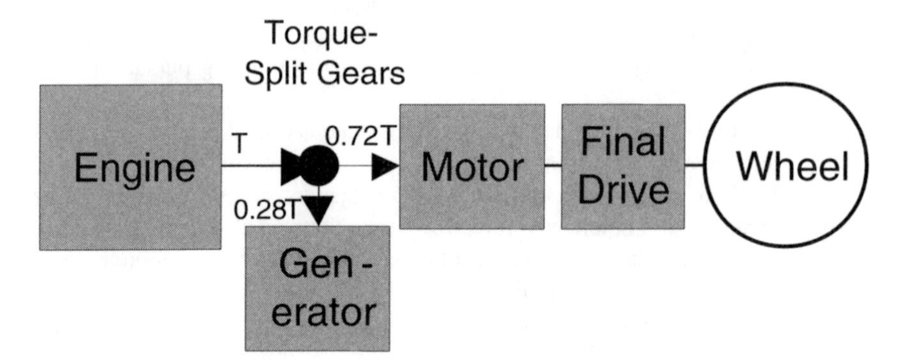

FIGURE 3-2 Toyota Prius power-train configuration. The components shown are the energy converters, not storage devices. The batteries are not shown.
SOURCE: Duoba et al., 2001.

al., 2001). Fuel-economy (gas-mileage) data were presented for three driving cycle tests: the Japanese 10-15 mode test, the Federal Test Procedure (FTP), and the highway test. Typically, in the United States a combined schedule of 55 percent urban and 45 percent highway fuel economies is used as the basis for vehicle gas-mileage comparison. Even though the reported results do not represent EPA-certified numbers, they are instructive for comparison purposes. The FTP fuel economy results were 2.07 gallons per 100 miles (48.2 mpg) and 1.65 gallons per 100 miles (60.7 mpg) for the Prius and Insight, respectively, somewhat lower than the results cited in the committee's sixth report (NRC, 2000). The difference presumably is caused by the variation in production vehicles. The FTP fuel economy results are impressive; however, it is also insightful to observe the differences in fuel economy of the two vehicles for the different types of driving, for example, highway driving versus city driving.

The Honda Insight is described by its manufacturer as having an integrated motor assist (IMA). The motor, integrated with a small engine, provides acceptable vehicle acceleration and recaptures energy during braking operations. The regenerative braking is simply done in parallel with the friction brake. In fact, if the driver disengages the clutch during braking, there is no regenerative braking. The Insight does not use the electric motor to produce power at low vehicle speeds or at low engine loads. The primary benefit of this hybridization scheme is reducing the engine size, thereby allowing it to operate more efficiently. As a result the driver benefits from excellent gas mileage during steady-speed driving, with good transient response supplied by the motor assist. Steady-speed driving tests of the Insight resulted in very low fuel consumption (high gas mileage).

In contrast to the Insight, the Toyota Prius uses a much more complicated power train and system control logic. Toyota uses a planetary gear system to couple the engine, generator, and motor in a way that provides a continuously variable transmission (AuYeung et al., 2001). This torque split is shown in Figure 3-2. In this configuration electrical power is required to keep the effective gear ratio fixed during operation at steady vehicle speed. This arrangement provides a great deal of flexibility. The engine can be turned off and on rapidly without using a clutch, electric power can be generated when the vehicle is at rest, and the vehicle can be driven in the electric-only mode. The regenerative braking scheme is significantly more complicated than that of the Insight. It is an active system that diminishes the friction braking depending on the power absorption capability of the electrical system and the driver-demanded deceleration rate to maximize the amount of energy recovery. This vehicle power train is optimized for both stop-and-go as well as low-speed driving. As part of the vehicle reconfiguration for introduction into the U.S. market, Toyota installed a larger engine to accommodate a higher percentage of highway driving, more typical in this country compared with Japan. When one considers the different approaches taken in hybridizing the two vehicles, it is perhaps not surprising that the Prius gets better gas mileage than the Insight on city-driving dynamometer tests, even though it is a larger and more massive vehicle (Duoba et al., 2001).

One approach to putting the comparison between the Honda Insight and the Toyota Prius in perspective is to describe the degree of hybridization of each vehicle. Decisions on the extent to which the fuel economy benefits of hybridization are to be incorporated into the vehicle's power train must be countered by assessments of the increased complexity, expense, and diminishing return of implementing the larger degree of hybrid technologies. From this perspective it becomes convenient to describe the approach to vehicle hybridization as a continuum, from no hybridization, to mild hybridization, to extensive hybridization. In this continuum the Honda Insight would be classified as a very mild hybrid and the Toyota Prius would be classified as having moderate hybridization.

How do these two vehicles compare with the hybridization effort being pursued by the PNGV partnership? There are similarities and some major differences. The PNGV concept cars are more similar to the Prius. In pursuit of maximum fuel economy, the concept cars incorporated all the design principles allowed by hybridization, namely, engine downsizing, battery-only electric driving at low loads, and regenerative braking. However, the extent to which the individual vehicles are hybridized and the control logic being pursued for each PNGV concept car are quite different. To meet the Goal 3 fuel economy target the PNGV hybrid vehicles will require a much larger degree of hybridization than has been demonstrated in either the Insight or the Pruis. The PNGV hybrid power trains and control systems are targeted for a different class of vehicle and a more lofty mileage target than those pursued by the Prius and Insight. The Toyota Prius and the Honda Insight are both smaller than the PNGV target vehicle; yet neither of them comes close to the PNGV target gas mileage of 80 mpg (1.25 gallons per 100 miles) for the FTP.

The USCAR partners face the same problem as both Honda and Toyota, in that they must overcome significant cost barriers in order to produce HEVs that have little or no cost penalty over today's single-power-plant vehicle. In this context the HEVs introduced by Toyota and Honda represent very interesting case studies. Both companies are gaining experience in manufacturing, marketing, and maintaining HEVs by subsidizing these vehicles and certainly must be looking for ways to migrate aspects of this technology into the rest of their vehicle fleet for competitive advantage. Consequently, the introduction of these hybrid vehicles is viewed as a significant occurrence and should be followed closely.

CONCEPT CARS AND PRODUCTION PROTOTYPES

The concept cars introduced in 2000 marked the completion of a major milestone for the PNGV program. Each company designed a car meeting its own vision of how to best approach the challenge of Goal 3 using both proprietary technology and technology developed in the cooperative program. Subsequent development of these vehicles has proceeded during 2000-2001. A concept car is the embodiment of a complete design using component part designs that represent final production intent but are not yet validated for production readiness. These concept cars have been used to validate simulation models, develop control systems, run performance tests, and evaluate driving characteristics. This activity generates knowledge from which components can be modified and a production-ready design can be developed. Because of its proprietary nature, most of this activity is carried out as a part of the PNGV but is not a partnership activity.

The next step called for in the PNGV Program Plan is the design, development, manufacture, and assembly of a production prototype by 2004. A production prototype is a car with components that have been validated as production ready, meaning that, at a minimum, a production process has been identified that is capable of manufacturing all the car's components in volume and with the required quality. This prototype car should also demonstrate all the characteristics required in order to make it an attractive, salable product. There should be a well-defined path to resolve any deficiencies from this standard.

At the time of the committee review the car companies were not ready to discuss their plans for moving from the concept-car stage into a production prototype. In proprietary discussions with the committee each company reviewed its plans to move major portions of the technology developed in the PNGV program into a variety of production vehicle programs. None of these vehicle programs was a simple extension of the power-train and vehicle design aspects of the concept cars.

As discussed in Chapter 4, "Program Overview," the committee believes that it is unrealistic to expect production prototypes to be built in the configurations of the concept cars. The cost and complexity of such vehicles, and the changed market environment since the program began, call into question the wisdom of following this original plan. Each of the companies is considering how to deal with this issue. The logical business decision is to apply derivative forms of the technology developed in the PNGV program to types of vehicles in which the increased costs may be better supported by market forces. This activity is exactly what was envisioned under Goal 2, but it leaves open the question of what course should be pursued under Goal 3. The committee recommends that the parties involved redefine this goal along the lines suggested in Chapter 4.

4

Program Overview

THE CHANGING CONTEXT

The purpose of the PNGV program is to improve substantially the fuel efficiency of today's automobiles and enhance the U.S. domestic automobile industry's productivity and competitiveness. The primary motivation for this effort came from a rising concern for the nation's rapidly increasing energy consumption, the threats to economic and political security from supply disruptions and price spikes in oil, and the increasing competition in the automotive industry from foreign companies.

Three specific goals of the PNGV followed directly from the overall purpose at the start of the partnership in 1993: (1) improvements in national competitiveness in the overall enterprise of manufacturing vehicles, (2) rapid implementation of innovations in conventional vehicles, and (3) development of marketable vehicles with up to three times the fuel efficiency of comparable 1994 family sedans.

However, since 1993 significant changes have occurred in the context surrounding the partnership. While not unpredicted, the U.S. automobile population has increased by 17 percent, the number of miles traveled has increased by 18 percent, the petroleum used in highway transportation has increased by 20 percent (DOT, 2000), and between 1995 and 2000 the percentage of the U.S. petroleum demand supplied by imports has increased from 44 percent to 52 percent (EIA, 1998). Also, the developing awareness of possible climate change from additions of greenhouse gases to the atmosphere has brought on a strong desire in some quarters to decrease carbon dioxide emissions to the atmosphere. These changes have greatly increased the incentive for the nation to conserve petroleum and added emphasis to the PNGV purpose to reduce passenger-car fuel consumption.

On the other hand, during this time the demand for sport utility vehicles (SUVs) and pickup trucks has increased to 46 percent of new light-duty vehicle sales. Due to their larger size and weight these vehicles have reduced the average new light-duty vehicle fleet fuel economy from 25.9 mpg in 1987 to 24 mpg in 2000 (EPA, 2000). There is also an implication that, if these lower-fuel-economy vehicles continue to dominate market share, the fleet average fuel economy will continue to decline. Also, the promulgation of Tier 2 emission requirements by the EPA potentially increases fuel consumption of all new cars and will postpone or possibly preclude the widespread use of diesel engines in light-duty vehicles in the United States.

During the past several years foreign and domestic competition in the automotive industry has become ever more difficult to define. Foreign companies (including DaimlerChrysler) produced about 45 percent of the new cars and light trucks sold in 1999 in the United States (U.S. Business Reporter, 2001), and GM and Ford both have added numerous overseas subsidiaries and collaborators to their corporate structures.

This new context for the PNGV makes it timely to consider whether the original goals should be altered to fit the changed situation while recognizing the substantial accomplishments already gained in pursuing the program so far.

PROGRAM PERSPECTIVES

Goal 1

Goal 1 relates to national competitiveness in manufacturing of vehicles. Improvements in manufacturing are clearly needed in order to make high-efficiency vehicles affordable and therefore marketable. In some cases manufacturing innovations are needed simply to make certain that new technologies are technically feasible. Thus, many PNGV activities in response to Goal 1 are easily justified as addressing critical barriers to success in Goal 3.

On the other hand, in today's context the broad issue of U.S. vehicle manufacturing competitiveness is very complex and unlikely to be seriously impacted by a program with the size and characteristics of the PNGV. The USCAR participants spend huge sums on developing manufacturing technologies as a part of their core business activity. It is true that these companies are in serious competition with foreign auto companies in the market, and manufacturing cost is an important element in that competition. But today all the USCAR participants are global marketers, all have foreign subsidiaries, one is a foreign-owned company, and there are even R&D collaborations with foreign companies aimed at the development and manufacture of major vehicle components (e.g., GM and

Toyota). On this basis it seems inappropriate to justify the broad PNGV manufacturing R&D charter as addressing U.S. competitiveness.

It also should be recognized that today foreign car companies market over 40 percent of the passenger cars sold in the United States. This fact, too, makes the Goal 1 competitiveness concept difficult to justify. In order to reduce the nation's petroleum fuel consumption it is desirable that all vehicles have reduced fuel consumption. For this to occur, the high-efficiency PNGV technologies and their associated manufacturing processes will have to be incorporated in cars sold in the United States by foreign manufacturers as well.

Notwithstanding these questions of scope and definition, the committee believes that the Goal 1 activities in the PNGV to date have been quite successful as detailed in earlier reports (NRC, 1999, 2000). Several manufacturing improvements actually have been implemented in production, and a number of others are well on their way. Many new projects to resolve some of the manufacturing challenges presented by the Goal 3 concept vehicles have been identified and ranked in cost/benefit order. It appears that there may be too few manufacturing people and financial resources in the program to pursue the bulk of these projects. Thus, the committee believes that the PNGV manufacturing goal should be tied closely to the fuel-efficiency objective of the program. Manufacturing R&D is vital to assure that the required new components can be produced with appropriate quality, cost, and predictability and can ultimately result in marketable vehicles. In today's context this focus seems more appropriate than a broad charter to impact overall U.S. vehicle manufacturing competitiveness.

Goal 2

Goal 2 requires that whatever desirable technologies are developed in the PNGV program and are ready for production be implemented in conventional vehicles as soon as possible without waiting to be applied in the production-prototype cars in 2004. The committee believes that there has been outstanding progress toward this goal. Several manufacturing and engineering innovations have already been put to use by the industry (NRC, 1999, 2000). Lightweight materials have found many new applications in current vehicles. And most importantly, as noted above, all three automobile companies are planning in true competitive fashion to introduce vehicles with HEV power trains and substantially reduced fuel consumption before the end of 2004.

Goal 3

The specific stretch goal to build a prototype, marketable, near-80-mpg mid-size car has served to provide the program with focus and a set of challenging but realistic research targets. The result has been the creation of industry-government R&D teams that have investigated a wide variety of potential technologies, selected and advanced the most critical of these, and enabled the construction of advanced, well-engineered concept cars in only six years. The automobile companies have invested many hundreds of millions of dollars in achieving this result, and this activity has led each of them to HEV product marketing plans.

General Motors has announced plans to produce an HEV Silverado pickup truck in 2004 and has indicated that its ParadiGM HEV system would be available worldwide across a variety of market segments from compacts to SUVs starting in 2004. DaimlerChrysler has announced that its Dodge Durango HEV will be introduced in 2003. Ford plans HEV Escape and Explorer vehicles in 2003. All these will be gasoline engine hybrids rather than diesel; they will not have fuel economies of 80 mpg, but will be 10 to 30 percent better than comparable current vehicles. As sales increase, these vehicles will have a substantial impact on fuel consumption in high-volume, high-use market segments.

In terms of the objective to decrease fleet fuel consumption, the production of these vehicles is especially significant. A 20 percent improvement in miles per gallon of a light-duty truck starting at 15 mpg would save about 155 gallons of fuel in a year, whereas the same percentage improvement of a mid-size sedan at 28 mpg saves only 83 gallons. Clearly, applying PNGV HEV technologies to light-duty trucks, vans, and SUVs will save more fuel per vehicle than applying them to mid-size sedans as currently specified in Goal 3.

As noted in Chapter 3, Goal 3 calls for a production-prototype mid-size sedan to be built by 2004. Early in the program it was decided that each of the automotive companies would build its own Goal 3 production-prototype vehicle, since this activity could not be performed in the context of precompetitive research. It seems clear now that an affordable mid-size sedan with three times the fuel economy of its 1994 counterpart cannot be a part of the foreseeable product marketing programs of any of the manufacturers at the present time. While performance, comfort, cargo space, utility, and safety targets can be met, the combination of 80 mpg and affordability appears well out of reach. Furthermore, it would be wasteful at this point to develop a production prototype for a vehicle that could not be marketed. The immense resources required to build and validate such a prototype can be justified only if the vehicle is part of the manufacturer's forward product marketing plan.

In light of the above facts, the committee concludes that it would be more appropriate for the program vehicle-efficiency goal to be based on fuel consumption and total fuel savings instead of miles per gallon. Fuel consumption more directly measures the resulting decrease in the use of petroleum-based fuels that would result if the improved vehicle-efficiency targets were achieved. If the program goal were refocused on reducing total new light-duty vehicle petroleum consumption, this would encourage emphasis to be placed on those vehicles that offer the greatest potential for achieving this societal goal. For better public understanding, the overall fuel consumption reduction could be translated into a goal of miles-per-gallon improvement for vehicles that start from various base fuel economy levels.

Events have shown that any restructuring and refocusing of the program should consider the merits and problems created by choosing a specific target vehicle type to use in measuring efficiency improvement. A specific vehicle target does provide a clear objective that can be simply stated. In the PNGV program to date the mid-size car target has resulted in concept vehicles that illustrate three different approaches to meeting conflicting program goals. However, choosing just one target vehicle type can also result in a narrower search for new technology and thus may be counterproductive. Clearly, the PNGV program intends that the promising marketable technology it develops be applied to *all* types of vehicles that can have a significant impact on total fleet fuel consumption.

Furthermore, the committee believes that it is inappropriate to include the process of building production prototypes in a precompetitive, cooperative industry-government program. The timing and construction of such vehicles is too intimately tied to the proprietary aspects of each company's core business to have this work scheduled and conducted as part of a joint, public activity. The success of the year 2000 concept cars suggests that perhaps second generation concept vehicles, still aimed at a stretch goal, might provide a suitable milestone in place of production prototypes.

SUMMARY

The overview and retrospective look at PNGV reported above led the committee to try to capture the essential factors that have created successes in this endeavor and note opportunities to enhance the future effectiveness of the program. In the committee's view the PNGV program has been a success largely because:

1. It set a deadline for a specific, focused, measurable stretch goal that was publicly visible, was judged to be important by government, and could motivate substantial industry support.
2. It required cooperation between industry and government organizations working on precompetitive research and development projects that were oriented toward the established goal, and it established an organizational structure that involved these key potential contributors in a truly collaborative manner.
3. It resulted in better use of the national laboratories and their tremendous resources to help resolve many of the very difficult issues.

The committee believes that, in the future, this activity would benefit substantially from:

1. A clear statement of the societal need being addressed and an exposition of the desired result that is subscribed to by the Administration, Congress, and industry partners;
2. Funding from the government at an overall level appropriate to the potential societal payoff, with the government-industry funding distribution set according to the risk level associated with individual projects;
3. Continuation of the program beyond the current 2004 termination date;
4. Deadlines appropriate for the tasks of major projects, subject to revision based on a rigorous, constructive review by the partners together with an independent outside body;
5. The construction of a series of concept vehicles and a review at agreed-on dates of technologies developed in the program that have been applied to production vehicles;
6. Creation of a high-level forum among the partners to discuss trade-offs among conflicting societal demands being addressed by the program and to recommend tactical and policy options; and
7. Emphasis on fundamental R&D for high-risk, high-payoff technologies.

The committee believes that, overall, the PNGV program is an excellent example of how long-range societal goals can be effectively addressed by the efforts of a collaborative, precompetitive government-industry R&D partnership. It is hoped that future activity will build on this program's success and enhance its effectiveness with lessons learned from the past several years.

Recommendation

Recommendation. Taking into consideration the successes, degree of progress, and lessons learned in the PNGV program to date, government and industry participants should refine the PNGV charter and goals to better reflect current societal needs and the ability of a cooperative, precompetitive R&D program to address these needs successfully.

5

PNGV's Response to the Sixth Report

In its previous six reviews, the National Research Council's Standing Committee to Review the Research Program of the PNGV made a number of recommendations, which are documented in published reports (NRC, 1994, 1996, 1997, 1998, 1999, 2000). In the sixth report the committee made specific recommendations related to each of the technologies under development and general recommendations for the program as a whole. Appendix B contains a letter from PNGV to the committee chair documenting PNGV's responses to the major recommendations in the Executive Summary of the sixth report (NRC, 2000). The committee believes that the PNGV has responded well to its recommendations and has been responsive to its suggestions. Discussions of PNGV's responses to the technical suggestions and recommendations in the sixth report are incorporated in the corresponding technical sections in Chapters 2 and 3.

Here, the committee simply makes a few points related to the PNGV responses. In recommendation 1[1] the committee suggested that PNGV should quantify and model the trade-off between efficiency and emissions for the power plants under consideration. Recommendation 11 also called for more systems modeling that could quantify the fuel-economy penalty associated with using different technologies to meet the new Tier 2 emission standards, including the effect of vehicle hybridization. PNGV's response was that predictive models do not have the accuracy to perform these trade-off studies or that they are too large, complex, and computationally intensive for use with software, such as Advisor or PSAT, developed for microcomputers. However, the committee has been encour-

[1]Recommendation numbers correspond to the recommendation numbers in Appendix B.

aged by the system modeling studies published recently by MIT and ANL (Weiss et al., 2000; Santini et al., 2001; An et al., 2001). These studies dealt strictly with fuel economies, greenhouse gases, and costs associated with various PNGV technologies, and not Tier 2 emission standards. Nevertheless, the committee believes that, with these studies as examples, the regulated emissions also can be dealt with in a useful, if rudimentary, manner. Even rudimentary modeling with engine steady-state emission maps and estimates of transient emissions (including cold starts and accelerations) would help to answer the question of whether hybridization controls can be adjusted away from the optimum fuel economy and toward reduced emissions.

Recommendation 8 called for a major study to determine how well lightweight PNGV vehicles would fare in collisions with the heavier vehicles that constitute the majority of the vehicle population entering service since 1995. As a partial response to safety issues, the Safety Working Group has been formed to identify and prioritize safety issues for PNGV-type vehicles. The committee believes that the formation of the Safety Working Group is a large step in the right direction. The Safety Working Group also recognized the high priority of the weight versus safety issue and has initiated statistical studies using existing accident databases to identify facts and trends pertaining to how vehicle size affects safety. However, the absence of PNGV-type vehicles in the database limits the ability to predict the safety of lightweight vehicles that are not small in size as well. The committee notes that no progress has been made on the fleet modeling study intended to complement the statistical analysis. In view of the difficulty of funding a study of the proposed magnitude, it is suggested that creativity be exercised by the Safety Working Group in defining and initiating affordable studies that would provide the most critically needed information that is lacking from the statistical study.

The issue of having an appropriate fuel for advanced PNGV vehicles has been a continuing issue that the committee has commented on in previous reviews. The committee recommended that the PNGV strengthen and expand its cooperative efforts with the petroleum industry. PNGV is exploring the possibility of having a joint symposium with the petroleum industry. It appears that relatively little progress has been made in response to this recommendation, but the committee understands that the petroleum and fuels industry is complex and that developing a good working relationship will take time and much effort on the part of the PNGV and the petroleum industry.

References

An, F., A. Vyas, J. Anderson, and D. Santini. 2001. Evaluating Commercial and Prototype HEVs. Society of Automotive Engineers (SAE) Paper No. 2001-01-0951, SAE World Congress, Detroit, Michigan, March 5-8. Warrendale, Pa.: Society of Automotive Engineers.

Anderman, M., F.R. Kalhammer, and D. MacArthur. 2000. Advanced Batteries for Electric Vehicles: An Assessment of Performance, Cost, and Availability (July). The Year 2000 Battery Technology Advisory Panel. Sacramento, Calif.: California Air Resources Board.

AuYeung, F., J.B. Heywood, and A. Schafer. 2001. Future Light Duty Vehicles: Predicting their Fuel Consumption and Carbon-Reduction Potential. SAE 2001-01-1081, SAE International Congress and Exposition, 2001 (March). Warrendale, Pa.: Society of Automotive Engineers.

CFR (Code of Federal Regulations). 2000. Part 49, Parts 400 to 999; Transportation; revised as of October 1, 2000; §571.111 Standard No. 111; Rear View Mirrors, pp. 326-335. Washington, D.C.: U.S. Government Printing Office.

DOE (U.S. Department of Energy). 1997. Scenarios of U.S. Carbon Reductions: Potential Impacts of Energy-Efficient and Low-Carbon Technologies to 2010 and Beyond. Washington, D.C.: U.S. Department of Energy, Interlaboratory Working Group on Energy-Efficient and Low-Carbon Technologies.

DOE. 2000a. Impact of Diesel Fuel Sulfur on CIDI Engine Emission Control Technology. (August). Washington, D.C.: U.S. Department of Energy, Office of Advanced Automotive Technologies.

DOE. 2000b. Research and Development and Analysis for Energy Efficient Technologies in Transportation and Buildings Applications. Solicitation issued by the Office of Energy Efficiency and Renewable Energy, November 21, 2000. Washington, D.C.: U.S. Department of Energy.

DOE. 2001. Project Abstracts for the U.S. Department of Energy Hydrogen Program, Annual Review Meeting, Baltimore, Md., April 17-19, 2001.

DOT (U.S. Department of Transportation). 1997. Relationship of Vehicle Weight to Fatality and Injury Risk in Model Year 1985-1993 Passenger Cars and Light Trucks, NHTSA Summary Report, DOT HS 808 569, April. Washington, D.C.: NHTSA.

DOT. 2000. Vehicle Registrations, Fuel Consumption, and Vehicle Miles of Travel as Indices (extrapolated to 2000). Washington, D.C.: Federal Highway Administration. Also available online at <http://search.bts.gov/ntl/query.html?qt=Vehicle+Registrations&search.x=11&search.y=3>.

Duoba, M., and H. Ng. 2001. Assessment of Japanese Hybrid Production Vehicles: Performance Issues. Presentation to the Standing Committee to Review the Research Program of the Partnership for a New Generation of Vehicles, USCAR Headquarters, Southfield, Mich., February 22, 2001.

Duoba, M., H. Ng, and R. Larsen. 2001. Characterization and Comparison of Two Hybrid Electric Vehicles (HEV's)—Honda Insight and Toyota Prius, SAE 2001-01-1335, SAE International Congress and Exposition, 2001 (March). Warrendale, Pa.: Society of Automotive Engineers.

EIA (Energy Information Administration). 1998. Annual Energy Outlook 1998 with Projections to 2020, Appendix A, Reference Case Forecast. Washington, D.C.: U.S. Department of Energy. Also available online at <http://www.eia.doe.gov/oiaf/aeo98/aeo98.html>.

EPA (U.S. Environmental Protection Agency). 2000. Light Duty Automotive Technology and Fuel Economy Trends 1975 Through 2000. EPA420-R-00-008 (December). Washington, D.C.: Environmental Protection Agency. Also available online at <http://www.epa.gov/otaq/fetrends.htm>.

Federal Register. 1999. Control of Air Pollution from New Motor Vehicles: Proposed Tier 2 Motor Vehicle Emissions Standards and Gasoline Sulfur Control Requirements—Regulatory Impact Analysis—May 13, 1999. Vol. 64, p. 92. Environmental Protection Agency Notice of Proposed Rule Making.

Federal Register. 2001. Control of Air Pollution from New Motor Vehicles: Heavy Duty Engine and Vehicle Standards and Highway Diesel Fuel Sulfur Requirements. January 18, 2001. Vol. 66, No. 12. Washington, D.C.: U.S. Government Printing Office.

Gardner, T.P., S.S. Low, and T.E. Kenny. 2001. Evaluation of Some Alternative Diesel Fuels for Low Emissions and Improved Fuel Economy. Society of Automotive Engineers (SAE) paper 2001-01-149, SAE International Congress and Exposition, March 2001. Warrendale, Pa.: Society of Automotive Engineers.

Hilden, D.L., J.C. Eckstrom, and L.R. Wolf. 2001. The Emission Performance of Oxygenated Diesel Fuels in a Prototype DI Diesel Engine. Society of Automotive Engineers (SAE) paper 2001-01-0650, SAE International Congress and Exposition, March 2001. Warrendale, Pa.: Society of Automotive Engineers.

Howden, K. 2000. 4SDI Technical Team Presentation. Presentation to the Standing Committee to Review the Research Program of the Partnership for a New Generation of Vehicles, USCAR Headquarters, Southfield, Mich., December 7, 2000.

Jeannes, D., and M. van Schaik. 2000. Steel Body Approaches to Reducing Vehicle Weight. Presentation to the Standing Committee to Review the Research Program of the Partnership for a New Generation of Vehicles, National Academy of Sciences, Washington, D.C., February 23, 2000.

Kenny, T.E., T.P. Gardner, S.S. Low, J.C. Eckstrom, L.R. Wolf, S.J. Korn, and P.G. Szymkowicz. 2001. Overall Results: Phase 1 Ad Hoc Diesel Fuel Test Program. Society of Automotive Engineers (SAE) paper 2001-01-0151, SAE International Congress and Exposition, March 2001. Warrendale, Pa.: Society of Automotive Engineers.

Korn, S.J. 2001. An Advanced Diesel Fuels Test Program. Society of Automotive Engineers (SAE) paper 2001-01-150, SAE International Congress and Exposition, March 2001. Warrendale, Pa.: Society of Automotive Engineers.

Mehta, M. 2000. Materials Technical Team Review. Presentation to the Standing Committee to Review the Research Program of the Partnership for a New Generation of Vehicles, USCAR Headquarters, Southfield, Mich., December 7, 2000.

Muller, J.T. 2000. Direct Methanol Fuel Cells: Paving the Way for a Clean Fuel. Daimler-Chrysler Innovation Symposium, November 8-9, 2000, Sindelfingen, Germany.

NRC (National Research Council). 1992. Automotive Fuel Economy. Washington, D.C.: National Academy Press.

NRC. 1994. Review of the Research Program of the Partnership for a New Generation of Vehicles. Washington, D.C.: National Academy Press.

NRC. 1996. Review of the Research Program of the Partnership for a New Generation of Vehicles, Second Report. Washington, D.C.: National Academy Press.

NRC. 1997. Review of the Research Program of the Partnership for a New Generation of Vehicles, Third Report. Washington, D.C.: National Academy Press.

NRC. 1998. Review of the Research Program of the Partnership for a New Generation of Vehicles, Fourth Report. Washington, D.C.: National Academy Press.

NRC. 1999. Review of the Research Program of the Partnership for a New Generation of Vehicles, Fifth Report. Washington, D.C.: National Academy Press.

NRC. 2000. Review of the Research Program of the Partnership for a New Generation of Vehicles, Sixth Report. Washington, D.C.: National Academy Press.

OTA (Office of Technology Assessment). 1995. Advanced Automotive Technology: Visions of a Super-Efficient Family Car. OTA-ETI-638. Washington, D.C.: U.S. Government Printing Office.

PCAST (President's Committee of Advisors on Science and Technology). 1997. Federal Energy Research and Development for the Challenges of the Twenty-First Century. November 5. Washington, D.C.: Executive Office of the President.

Plotkin, S. 2001. European and Japanese Initiatives to Boost Automotive Fuel Economy—What They Are, Their Prospects for Success, Their Usefulness as a Guide for U.S. Actions. Presentation to the 80th Annual Meeting, Transportation Research Board, January 7-11, 2001.

PNGV (Partnership for a New Generation of Vehicles). 1995. Partnership for a New Generation of Vehicles Program Plan (draft). Washington, D.C.: U.S. Department of Commerce, PNGV Secretariat.

PNGV. 1999. Technical Roadmap (updated draft, September, 1999). Southfield, Mich.: PNGV/USCAR.

PNGV. 2001. Written Response of PNGV Executive Committee to Questions Raised by Standing Committee, February 2.

Powers, W.F. 2000. Automotive Materials in the 21st Century. In Advanced Materials and Processes, p. 38. Materials Park, Oh.: ASM International.

Santini, D., A. Vyas, J. Anderson, and F. An. 2001. Estimating Trade-offs Along the Path to the PNGV 3X Goal. Transportation Research Board (TRB) Paper 01-3222. TRB 80th Annual Meeting, Washington, D.C., January 7-11, 2001.

Schultz, R.A. 1999. Aluminum for Lightweight Vehicles: An Objective Look at the Next 10 Years to 20 Years. Presented at the Metal Bulletin 14th International Aluminum Conference, Montreal, Canada, September 15, 1999. Available from Richard A. Schultz, Ducker Research Co., Inc. (<richards@drucker.com>).

Sissine, F. 1996. The Partnership for a New Generation of Vehicles (PNGV). Report No. 96-191 SPR. Washington, D.C.: Congressional Research Service.

Stuef, B. 1997. Vehicle Engineering Team Review. Presentation to the Standing Committee to Review the Research Program of the Partnership for a New Generation of Vehicles, USCAR Headquarters, Southfield, Mich., October 14, 1997.

Szymkowski, P.G., D.T. French, and C.C. Crellin. 2001. Effects of Advanced Fuels on the Particulate and NO_x Emissions from an Optimized Light-duty CIDI Engine. 2001. Society of Automotive Engineers (SAE) paper 2001-01-148, SAE International Congress and Exposition, March 2001. Warrendale, Pa.: Society of Automotive Engineers.

Toyota. 2000. International Workshop on Next Generation Power Systems. Tokyo, Japan, September.

ULSAB (UltraLight Steel Auto Body). 1999. Available online at <http:/www.ulsab.org>.

U.S. Business Reporter. 2001. The Digital Research Tool of Business, Automobile Industry Market Shares (March). Also available online at <http://www.activemedia-guide.com/mrkid_automobile.htm>.

Wallace, J. 2001. Update on California Fuel Cell Partnership. Presentation to the Standing Committee to Review the Research Program of the Partnership for a New Generation of Vehicles. USCAR Headquarters, Southfield, Mich., February 22, 2001.

Weiss, M., J.B. Heywood, E.M. Drake, A. Schafer, and F.F. AuYeung. 2000. On the Road in 2020, a Life-Cycle Analysis of New Automobile Technologies. Energy Laboratory, Massachusetts Institute of Technology (October), Cambridge, Mass.

The White House. 1993. Historic Partnership Forged with Automakers Aims for Threefold Increase in Fuel Efficiency in as Soon as Ten Years. Washington, D.C.: The White House.

Appendixes

Appendix A

Biographical Sketches of Committee Members

Craig Marks, chair, retired as vice president of technology and productivity for AlliedSignal Automotive, where he was responsible for product development; manufacturing; quality; health, safety, and environment; and communications and business planning. Dr. Marks is now chairman of the Board of Trustees of the Environmental Research Institute of Michigan. After retiring, he was adjunct professor in both the College of Engineering and the School of Business Administration at the University of Michigan and codirector of the Tauber Manufacturing Institute. Previously, in TRW's Automotive Worldwide Sector, Dr. Marks was vice president for engineering and technology and later vice president of technology at TRW Safety Restraint Systems. Prior to joining TRW, he held various positions at GM Corporation, including executive director of the engineering staff; assistant director of advanced product engineering; engineer in charge of power development; electric-vehicle program manager; supervisor for long-range engine development; and executive director of the environmental activities staff. He is a member of the National Academy of Engineering and a fellow of the Society of Automotive Engineers. Dr. Marks received his B.S.M.E., M.S.M.E., and Ph.D. in mechanical engineering from the California Institute of Technology.

Vernon P. Roan, vice chair, is director of the Center for Advanced Studies in Engineering and professor of mechanical engineering at the University of Florida, where he has been a faculty member for more than 30 years. Previously, he was director of the Center for Advanced Studies in Engineering and, since 1994, he has also been the director of the University of Florida Fuel Cell Research and Training Laboratory. Previously, he was a senior design engineer with Pratt and Whitney Aircraft. Dr. Roan, who has more than 25 years of research and development experience, is currently developing improved modeling and simulation systems for a fuel cell bus program and working as a consultant to Pratt and Whitney on advanced gas-turbine propulsion systems. His research at the University of Florida has involved both spark-ignition and diesel engines operating with many alternative fuels and advanced concepts. With groups of engineering students, he designed and built a 20-passenger diesel-electric bus for the Florida Department of Transportation and a hybrid-electric urban car using an internal-combustion engine and lead-acid batteries. He has been a consultant to the Jet Propulsion Laboratory monitoring their electric and hybrid-vehicle programs. He has organized and chaired two national meetings on advanced vehicle technologies and a national seminar on the development of fuel-cell-powered automobiles and has published numerous technical papers on innovative propulsion systems. He was one of the four members of the Fuel Cell Technical Advisory Panel of the California Air Resources Board, which issued a report in May 1998 regarding the status and outlook for fuel cells for transportation applications. Dr. Roan received his B.S. in aeronautical engineering and his M.S. in engineering from the University of Florida and his Ph.D. in engineering from the University of Illinois.

William Agnew retired as director, Programs and Plans, General Motors Research Laboratories in 1989. He served in the Manhattan District from 1944 to 1946, and attended Purdue University from 1946 to 1952. From 1952 to 1989 he held a number of positions at GM Research Laboratories, including department head, Fuels and Lubricants; head, Emissions Research Department; technical director, Engine Research, Engineering Mechanics, Mechanical Research, Fluid Dynamics, and Fuels and Lubricants Departments; technical director, Biomedical Science, Environmental Science, Societal Analysis, and Transportation Research Departments. Dr. Agnew's technical expertise spans internal combustion engines, gas turbines, engine performance, automotive air pollution, and automotive power plants. A member of the National Academy of Engineering, he has a Ph.D. in mechanical engineering from Purdue University.

Kennerly H. Digges is research professor of engineering and director of biomechanics and automotive safety research, George Washington University. Previously, he was a senior executive in the U.S. Department of Transportation's National Highway Traffic Safety Administration (NHTSA). At NHTSA, he was head of the Research and Development Office and managed research to advance motor-vehicle crash-safety standards, such as side-impact protection, and led the development of experimental automobiles that protect occupants in severe crashes. He was also head of NHTSA's Rulemaking Office and contributed to the introduction of automatic restraints for new cars. Prior to joining NHTSA, Dr. Digges spent 10 years directing the U.S. Air Force research program in mechanical systems for aircraft. He is a past director of the Transportation Rehabilitation Engineering Center at the University of Virginia for the National Institute for

Disability and Rehabilitation Research. Dr. Digges is a member of the Society of Automotive Engineers, serving as a Technical Board member and a seminar instructor in computer accident reconstruction. He is an internationally recognized expert in the field of automotive safety and has published numerous papers on accident characterization and safety performance. He received his B.S. in mechanical engineering from Virginia Polytechnic Institute and his M.S. and Ph.D. in mechanical engineering from Ohio State University.

W. Robert Epperly is president of Epperly Associates, Inc., a consulting firm. From 1993 to 1997 he was vice president of Catalytica, Inc., and from 1995 to 1997 was president of Catalytica Advanced Technologies, Inc., a company that develops new catalytic technologies for the petroleum and chemical industries. Prior to joining Catalytica, he was general manager of Exxon Corporate Research and director of the Exxon Fuels Research Laboratory. After leaving Exxon he was chief executive officer of Fuel Tech N.V., a company that develops new combustion and air pollution control technologies. Mr. Epperly has written or coauthored more than 50 publications on technical and managerial topics, including two books, and has 38 U.S. patents. He has extensive experience in fuels, fuel cells, engines, catalysis, air pollution control, and the management of research and development programs. He received an M.S. degree in chemical engineering from Virginia Polytechnic Institute.

David E. Foster is professor of mechanical engineering, University of Wisconsin, Madison, and former director of the Engine Research Center, which has won two center of excellence competitions for engine research and has extensive facilities for research on internal combustion engines. A member of the faculty at the University of Wisconsin since he completed his Ph.D., Dr. Foster teaches and conducts research in thermodynamics, fluid mechanics, internal combustion engines, and emission formation processes. His work has focused on perfecting the application of optical diagnostics in engine systems and the incorporation of simplified or phenomenological models of emission formation processes into engineering simulations. He has published more than 60 technical articles in this field throughout the world and for leading societies in this country. He is a recipient of the Ralph R. Teetor Award, the Forest R. McFarland Award, and the Lloyd L. Withrow Distinguished Speaker Award of the Society of Automotive Engineers (SAE), and he is an SAE fellow. He is a registered professional engineer in the State of Wisconsin and has won departmental, engineering society, and university awards for his classroom teaching. He received a B.S. and M.S. in mechanical engineering from the University of Wisconsin and a Ph.D. in mechanical engineering from the Massachusetts Institute of Technology.

Norman A. Gjostein is a clinical professor of engineering at the University of Michigan, Dearborn, where he teaches courses in materials engineering. He retired from Ford Research Laboratory as director, Manufacturing and Materials Research Laboratory, which includes research in advanced materials, manufacturing systems, and computer-aided engineering. He has directed a variety of advanced research programs, including the development of lightweight metals, composite materials, sodium-sulfur batteries, fiber-optic multiplex systems, and smart sensors. He has pioneered studies in surface science and discovered a number of new surface structures that are still under investigation. He is a member of the National Academy of Engineering, a fellow of the Engineering Society of Detroit (ESD) and the American Society of Metals (ASM), and a recipient of the ASM's Shoemaker Award and ESD's Gold Award. Dr. Gjostein has a B.S. and M.S. in metallurgical engineering from the Illinois Institute of Technology and a Ph.D. in metallurgical engineering from Carnegie Mellon University.

David F. Hagen spent 35 years with Ford Motor Company, where his position prior to retirement was general manager, Alpha Simultaneous Engineering, Ford Technical Affairs. Under his leadership Ford's alpha activity, which involves the identification, assessment, and implementation of new product and process technologies, evolved into the company's global resource for leading-edge automotive products, processes, and analytic technologies. Mr. Hagen led the introduction of the first domestic industry feedback electronics, central fuel metering, full electronic engine controls, and numerous four-cylinder, V6, and V8 engines. Based on his work on Ford's modern engine families, he was awarded the Society of Automotive Engineers E.N. Cole Award for Automotive Engineering Innovation in 1998. Mr. Hagen was employed by the Michigan Center for High Technology from 1995 to 1997 and was its president in 1997. He received his B.S. and M.S. in mechanical engineering from the University of Michigan. He is a fellow of the Engineering Society of Detroit and its past president and a member of the Society of Automotive Engineers. He is currently serving on the Engineering Advisory Board of the University of Michigan, Dearborn.

John B. Heywood is Sun Jae Professor of Mechanical Engineering at the Massachusetts Institute of Technology and director of the Sloan Automotive Laboratory. Dr. Heywood's research interests have focused on understanding and explaining the processes that govern the operation and design of internal combustion engines and their fuel requirements. His major areas of research include engine combustion, pollutant formation, operating and emissions characteristics, and fuel requirements of automotive and aircraft engines. He has been a consultant to Ford Motor Company, Mobil Research and Development Corporation, and several other industry and government organizations. He received the U.S. Department of Transportation 1996 Award for the Advancement of Motor Vehicle Research and Development, and several awards from the Society of Automotive Engineers (SAE) and the American Society of Mechanical Engineers. He is a fellow of SAE and a member of the National Academy of Engineering. He has a Ph.D. in mechanical engineering from the Massachusetts Institute of Technology, an Sc.D. from Cambridge University, and an honorary D. Tech from Chalmers University of Technology.

Fritz Kalhammer is a consultant for the Electric Power Research Institute's (EPRI) Strategic Science and Technology and Transportation Groups. He was cochair of the California Air Resources Board's Battery Technical Advisory Panels on electric vehicle batteries, and he recently chaired a similar panel to assess the prospects of fuel cells for electric vehicle propulsion. He has been vice president of EPRI's Strategic Research and Development and established the institute's programs for energy storage, fuel cells, electric vehicles, and energy conservation. Before joining EPRI, he directed electrochemical energy conversion, storage, and process research and development at Stanford Research Institute (now SRI International), conducted research in solid-state physics at Philco Corporation, and conducted research in catalysis at Hoechst in Germany. He recently completed an assignment as chair of the Year 2000 Battery Technical Advisory Panel to the California Air Resources Board, with a final report issued by the panel. He has a Ph.D. in physical chemistry from the University of Munich.

John G. Kassakian is professor of electrical engineering and director of the Massachusetts Institute of Technology (MIT) Laboratory for Electromagnetic and Electronic Systems. His expertise is in the use of electronics for the control and conversion of electrical energy, industrial and utility applications of power electronics, electronic manufacturing technologies, and automotive electrical and electronic systems. Prior to joining the MIT faculty he served in the U.S. Navy. Dr. Kassakian is on the boards of directors of a number of companies and has held numerous positions with the Institute of Electrical and Electronics Engineers (IEEE), including founding president of the IEEE Power Electronics Society. He is a member of the National Academy of Engineering, a fellow of the IEEE, and a recipient of the IEEE's William E. Newell Award for Outstanding Achievements in Power Electronics, the IEEE Centennial Medal, and the IEEE Power Electronics Society's Distinguished Service Award. He has an Sc.D. in electrical engineering from MIT.

Harold H. Kung is professor of chemical engineering at Northwestern University. His areas of research include surface chemistry, catalysis, and chemical reaction engineering. His professional experience includes work as a research chemist at E.I. du Pont de Nemours & Co. He is a recipient of the P.H. Emmett Award and the Robert Burwell Lectureship Award from the North American Catalysis Society, the Herman Pines Award of the Chicago Catalysis Club, the Japanese Society for the Promotion of Science Fellowship, the John McClanahan Henske Distinguished Lectureship of Yale University, and the Olaf A. Hougen

Professorship at the University of Wisconsin, Madison. He has a Ph.D. in chemistry from Northwestern University.

David F. Merrion is a retired executive vice president, Engineering, for Detroit Diesel Corporation. His positions at Detroit Diesel included staff engineer, Emissions and Combustion; staff engineer, Research and Development; chief engineer, Applications; director, Diesel Engineering; general director, Engineering (Engines and Transmissions); and senior vice president, Engineering. He has extensive expertise in the research, development, and manufacturing of advanced diesel engines, including alternative fuel engines. He is a Society of Automotive Engineers (SAE) fellow and a member of American Society of Mechanical Engineers. He served as a former president of the Engine Manufacturers Association; member of EPA's Mobile Sources Technical Advisory Committee; member of the Coordinating Research Council; and member of the U.S. Alternative Fuels Council. He has a bachelor of mechanical engineering from General Motors Institute/Kettering University and a master of science in mechanical engineering from the Massachusetts Institute of Technology.

John Scott Newman is a professor of chemical engineering at the University of California, Berkeley. His research experience is in the design and analysis of electrochemical systems, transport properties of concentrated electrolytic solutions, and various fuel cells and batteries. He has received the Young Author's Prize from the Electrochemical Society, the David C. Grahame Award, the Henry B. Linford Award, and the Olin Palladium Medal. He is a member of the National Academy of Engineering and a fellow of the Electrochemical Society. He is author of *Electrochemical Systems* (Prentice Hall, 1991), which has been translated into Japanese and Russian, and was an associate editor of the *Journal of the Electrochemical Society* from 1990 to 2000. He has a Ph.D. in chemical engineering from the University of California, Berkeley.

Roberta Nichols is retired from Ford Motor Company, where from 1979 to 1995 she held several positions, including manager, Electric Vehicle External Strategy and Planning Department, North American Automotive Operations; manager, EV External Affairs, EV Planning and Program Office; manager, Alternative Fuels Department, Environment and Safety Engineering Staff; and principal research engineer, Alternative Fuels Department, Scientific Research Laboratory. She was also a member of the technical staff of Aerospace Corporation from 1960 to 1979 and has held other industry positions. She is a fellow of the Society of Automotive Engineers, a recipient of the National Achievement Award of the Society of Women Engineers, a recipient of the Clean Air Award for Advancing Air Pollution Technology of the South Coast Air Quality Management District, and a member of the National Academy of Engineering. Her expertise includes alternative fuel vehicles, electric vehicles, internal combustion engines, and strategic planning. She has a Ph.D. in engineering and an M.S. in environmental engineering from the University of Southern California and a B.S. in physics from the University of California, Los Angeles.

F. Stan Settles is a professor and chair of industrial and systems engineering at the University of Southern California. Prior to his current role he served as program director for design and integration engineering at the National Science Foundation (NSF). Dr. Settles was on loan to the NSF from Arizona State University in Tempe, where he was a research professor in the Department of Industrial and Management Systems Engineering. In 1992 and early 1993 he served as assistant director for industrial technology in the White House Office of Science and Technology Policy. Dr. Settles had a 30-year career with Garrett/AlliedSignal (now Honeywell), primarily in Phoenix, Arizona. He held a number of positions in design and project engineering, manufacturing, and general management including manager of Industrial and Manufacturing Engineering, division director of Planning, division vice president of Manufacturing Operations, and corporate director of Industrial and Manufacturing Engineering. He is a fellow and past president of the Institute of Industrial Engineers, a senior member of the Society of Manufacturing Engineers, a member of the American Society for Quality, the American Society for Engineering Education, and the Institute for Operations Research and Management Science. His teaching and research interests are in the areas of quality management, engineering project management, and manufacturing systems engineering. He is a member of the National Academy of Engineering and served as the chair of the National Research Council's Board on Manufacturing and Engineering Design from 1994 to 1999. He has an M.S.E. and a Ph.D. in industrial engineering from Arizona State University. He holds B.S. degrees in both industrial engineering and production technology from LeTourneau University.

Appendix B

The PNGV Response to Recommendations in the Sixth Report

Partnership for a New Generation of Vehicles

November 29, 2000

Mr. Craig Marks
Chair, Standing Committee to Review the
Research Program of PNGV
National Research Council
2101 Constitution Avenue, N.W.
Room # HA 270
Washington, DC 20418

Dear Craig,

We want to thank you and the other Standing Committee members for your insightful and productive review of the Partnership for a New Generation of Vehicles.

Attached are our comments on the major recommendations from the 6th report. We agreed with many of your recommendations and we will be ready to discuss these with you at the upcoming 7th Review.

You also made a number of recommendations concerning the individual technology areas. The PNGV Technical Teams will address those areas during their discussions with you at the upcoming 7th Review.

Again we appreciate receiving your valuable analysis as we progress through the challenges of developing the PNGV technologies and advancing toward our goals.

Sincerely,

Vince Fazio
PNGV Director
Ford

Claude C. Gravatt
PNGV Secretariat
U.S. Department of Commerce

Ron York
PNGV Director
General Motors

Steve Zimmer
PNGV Director
Daimler-Chrysler

Attachment

Partnership For a New Generation of Vehicles

Response to the NRC's 6th Peer Review Report Recommendations

1) **RECOMMENDATION:** *PNGV should quantify the trade-off between efficiency and emissions for the power plants under consideration. The PNGV systems-analysis team should attempt to develop and validate vehicle emissions models of sufficient sophistication to provide useful predictions of the emissions potential for a variety of engines (e.g., the compression-ignition direct-injection engine, the gasoline direct-injection engine) and exhaust gas after-treatment systems in various hybrid vehicle configurations. The models could be used to help PNGV evaluate the feasibility of meeting the Environmental Protection Agency's Tier 2 emissions standards with various vehicle system configurations. These data should then be used to establish an appropriate plan for the next phase of the program.*

   RESPONSE: While such a modeling effort could be useful, we are unaware of any existing models that accurately predict emissions for multi-cylinder engines based on physical and chemical parameter inputs. Developing such a model would be a major undertaking with little guarantee of success.

2) **RECOMMENDATION:** *At this stage, PNGV should direct its program toward an appropriate compromise between fuel economy and cost using the best available technology to ensure that a market-acceptable production-prototype vehicle can be achieved by 2004 that meets Tier 2 emission standards.*

   RESPONSE: Cost issues will be discussed during the Peer Review, Dec. 7-8, 2000.

3) **RECOMMENDATION:** *Given the potential of fuel-cell technology for meeting the efficiency and emissions objectives of the PNGV program, the systems-analysis team should increase its efforts to develop more complete and accurate fuel-cell system and component models to support the development and assessment of fuel-cell technology.*

   RESPONSE:

   1.) Both ADVISOR and PSAT simulation tools have been connected to GCTool (a detailed modeling and simulation tool for fuel cell system). GCtool is used to update the model and the data in both ADVISOR and PSAT.

   2.) By working with Virginia Tech. University, the fuel cell future car has been used to validate ADVISOR model.

3.) A study of the effect of degree of hybridization on fuel economy for a fuel cell vehicle is in process.

4.) More detailed validation of the fuel cell model, that includes transients and warm-up effects, is in process with Virginia Tech.

4) **RECOMMENDATION:** *In the area of fuel cell development, PNGV, and especially the U.S. Department of Energy, should emphasize high-risk, high-payoff research in critical areas, such as fuel processing, carbon monoxide-tolerant electrodes, and air-management systems.*

RESPONSE: We concur with Recommendation #4. The Department of Energy program is focused on the critical barriers addressed above, as well as other critical areas such as cathode improvements to increase cell voltage and stack efficiency, and high temperature membranes that will facilitate heat rejection and improve CO tolerance. In fuel processing R & D, the focus is on lowering the thermal mass to improve start-up time and efficiency, and on durable, high activity, shift catalysts to reduce the size, weight, and cost of the fuel processor. Los Alamos is continuing to work on CO-tolerant anodes. In the air management area, a peer review of compressor/expander technology was conducted during 2000. Recommendations to reevaluate PNGV air compressor technical requirements, and to downselect to focus on the most promising compressor technology, will be implemented in 2000 and 2001, respectively.

5) **RECOMMENDATION:** *PNGV should continue to work on cell chemistry of lithium battery systems to extend life and improve safety, while continuing to lower costs. Performance and cost targets should be refined as overall vehicle systems analysis determines the optimal degree of vehicle hybridization.*

RESPONSE: As recommended by the Committee, lithium battery chemistry development aimed at extending life, improving safety and reducing cost, is continuing. The PNGV battery cost and performance targets have been refined periodically, based on (informal) direction from the proprietary vehicle development efforts. PNGV is considering the development of collaborative vehicle-level cost, performance, and fuel economy models. It may be possible to determine more optimal hybrid vehicle configurations from such models, and the battery performance and cost targets could then be further refined.

6) **RECOMMENDATION:** *As the PNGV program moves toward the 2004 production-prototype milestone, affordability will be a key requirement. Therefore, the development of an efficiently designed and fabricated steel-intensive vehicle being worked on by the American Iron and Steel Institute in Ultralight Steel Autobody-Advanced Vehicle Concept (ULSAB-AVC) project should be closely followed, and the possibility of applying the ULSAB concepts to a hybrid steel-aluminum vehicle should be explored.*

RESPONSE: The Materials Team (as well as the Vehicle Engineering and Manufacturing Teams) continue to monitor the progress of the AISI ULSAB-AVC and related projects. The MTT arranged for representatives of AISI to meet with MTT, VE, and Mfg. Team representatives on May 25 to review the status and plans for the

ULSAB-AVC program, and the results and findings of the ULSAC (Ultra-Light Steel Auto Closures) program. At that time, PNGV TEAM members offered feedback and suggestions to make the studies more pertinent / applicable to PNGV. It was agreed that PNGV would not be directly involved in the AISI programs, but AISI would continue to provide periodic updates to pertinent PNGV teams.

The MTT is also working with the Auto-Steel Partnership, in conjunction with USAMP, in formulating high strength steel projects (which build on the results of ULSAB projects) for possible inclusion in the portfolio of projects funded by DOE's OTT Lightweight Materials budget (possibly through USAMP/DOE Cooperative Agreement).

Implementation of lightweight steel techniques and technologies is primarily a function of the individual OEMs. There are no plans to jointly build a hybrid steel-aluminum vehicle. However, the MTT is supporting DOE funding of projects to be proposed by a recently formed USAMP Joining Task Force, which was specifically chartered to identify and address issues surrounding joining a mixture of low mass materials, such as high strength steels, aluminum, polymer composites, and magnesium.

7) **RECOMMENDATION:** *The committee recognizes the cost reduction potential of DaimlerChrysler's thermoplastic composite injection-molding technology and urges that this work be continued to bring the technology to successful commercialization. The committee encourages the earliest possible generation of vehicle and component test data to define better the structural properties and performance of various composite materials and structures.*

RESPONSE: The thermoplastic body development underway at DaimlerChrysler is proprietary to them. It will be deployed, when ready, in their products. The first announced product is the Jeep Wrangler hardtop scheduled for production in the Fall of 2001.

8) **RECOMMENDATION:** *The committee regards structural crashworthiness, and safety in general, in the design of lightweight PNGV vehicles as extremely important. Using the Oak Ridge National Laboratory car-to-car collision simulation capability, the National Highway Traffic and Safety Administration should support a major study to determine how well lightweight PNGV vehicles would fare in collisions with heavier vehicles and to assess potential improvements.*

RESPONSE: The Safety Working Group (SWG) agrees with the essence of the Committee's recommendation that structural crashworthiness and safety in general is critical for PNGV vehicles. The SWG was chartered in December of 1999 to identify challenges and technical issues that are relevant to the safety of vehicles developed under PNGV, and provide advice and analysis through the PNGV Executive Committee to the OSG on the safety of such vehicles and how to enhance their safety. The SWG is a team of safety experts from government and industry working together to address the safety needs for PNGV.

The top two of the highest priority research needs are being pursued for projects in the next year.

- **Size independent of weight**
  - two proposed studies
- **FMVSS 111 side view systems in lieu of mirrors**
  - one proposed study

One of the top priorities, "**Size independent of weight,**" is partially addressed by NHTSA's ongoing fleet modeling study. The completion of this study is still approximately one year away at $3.5M. The current FY 2001 budget does not include the funds to complete the fleet modeling study. NHTSA is searching for a low level of funding to continue the efforts at a slower pace. The Oak Ridge National Laboratory (ORNL) simulation efforts are part of this fleet modeling study. However, four other contractors in addition to ORNL are working on the fleet modeling study and they all have capabilities to perform the car-to-car simulations. NHTSA has not yet chosen a contractor to perform the car-to-car simulations.

In addition to the NHTSA fleet modeling study, the SWG is pursuing a statistical study based on existing accident databases to identify facts and trends pertaining to how size affects vehicle safety. This study would complement the fleet modeling study to determine the safety issues surrounding a lightweight but not small size PNGV vehicle.

The other top priority research study will address human factor issues with replacing planar driver side view mirrors with alternative side view systems. Since alternative side view systems would be a significant change in the manner in which drivers receive information about their external driving environment, steps must be taken to insure that such a system is robust, easily used, and acceptable to drivers.

9) **RECOMMENDATION:** *Defining automotive system/fuels trade-offs and establishing a basis for ensuring that the required fuels are available as higher efficiency vehicles become commercially available will require extensive cooperation among automotive and petroleum industry representatives at all levels of responsibility. Therefore, PNGV should strengthen and expand its cooperative efforts with the petroleum industry, including issues related to fuels for fuel cells. Government leadership will be necessary to initiate this cooperative effort and provide incentives for petroleum company involvement.*

RESPONSE: We are having discussions around the possibility of having a joint symposium with the petroleum industry.

10) **RECOMMENDATION:** *PNGV should consider conducting a comprehensive assessment of the consequences of fuel choices for fuel cells and their impact on PNGV's direction and ultimate goals. PNGV should assess the opportunities and costs for generating hydrogen for fuel cells at existing service stations and storing it on board vehicles and compare the feasibility, efficiency, and safety of this option with onboard fuel reforming. This study would help PNGV determine how much additional effort should be devoted to the development of onboard fuel reforming technologies.*

RESPONSE: We concur with Recommendation #10. We have initiated a feasibility study to assess cost, efficiency, and emissions (including CO_2) for various fuels/feedstocks, including off-board generation of hydrogen at existing refueling stations. We are also planning a solicitation in Fiscal Year 2001 to contract for research and development of low-cost off-board hydrogen generation, purification, storage, compression, and dispensing. The solicitation will also include on-board storage research to complement the off-board R & D. These combined efforts will result in an integrated demonstration of on-board storage and off-board refueling technologies.

11) **RECOMMENDATION:** *PNGV should continue its aggressive pursuit of lean combustion exhaust-gas after-treatment systems. The program should also pursue detailed systems modeling that could quantify the fuel economy penalty associated with using different technologies to meet the new Tier 2 emission standards. The modeling should also address how power train hybridization could be used to reduce emissions and what effects changing the primary energy converter would have on fuel economy trade-offs necessary to meet emissions standards.*

RESPONSE:
Background on emissions modeling:

- Experience has shown models that produce adequate results are much too large, complex and computational intensive to be of use in PC based models like Advisor and PSAT.
- In the past, models that would be appropriate for inclusion in PC based models would use steady state engine maps. The failure of these models to properly account for transient and cold-start effects was their weakness. Results using these models were not of sufficient quality to merit their use, except possibly in providing the most rudimentary, at best, of directional effects.
- Work on trying to address this issue is proceeding, as outlined below.
- However, the time required properly develop better models, select appropriate replacements and validate/calibrate these models will likely produce results, assuming success, that would be beyond the time framework that would make them useful to the current PNGV program.

Emissions modeling work at ORNL in FY 2000:

- A methodology is being developed to produce emission control device models based on analysis and reduction of emission data. Models for a catalyzed diesel particulates filter and a diesel oxidation catalyst are expected to be completed 3Q CY 2000.
- Mapping, including cold start data, of the 1.9L VW & 1.7L Mercedes CIDI with high pressure common rail has been completed.
- In order to be able to accurately measure ever decreasing emission levels, a new fast flow exhaust measurement capability is being developed which will also aid the understanding of transient emission behavior. Expected completion is 3Q CY2000.
- An engine scaling algorithm is being developed for Advisor, similar in nature to that already in PSAT with completion in FY 2001.

Emissions modeling work at ANL in FY 2000:

- Mapping of the 1.2L Ford Diata, 2.0L PSA & 1.7L Mercedes CIDI with high pressure common rail is beginning in CY 2000.
- The Modal Emission Model has been developed and is available as a candidate for possible use with if not inclusion in PSAT and Advisor.
- A new neural net emissions model is being developed specifically to address the need to account for transient/cold start emissions. Sufficient test data is being generated to define the model structures and "train" them. Preliminary results on Prius engine expected 3Q 2000 CY.

Emissions modeling work at ORNL in FY 2001:

- Work on methodologies to produce emission control device models based on analysis and reduction of emission data continues with the development of a NOX absorber model.
- Work will begin on developing an algorithm for determining the "state of warmness" of an engine for predicting emissions during cold start.
- Addressing the need for emissions data representative of SUV size, engine mapping will be done on a 5.9L Cummins DI diesel both, with and without EGR.

Emissions modeling work at ANL in FY 2001:

- Mapping of the 1.2L Ford Diata, 2.0L PSA & 1.7L Mercedes CIDI with high pressure common rail continues.
- The neural net emissions model will be further tested and, if successful, its use will be expanded.

The Systems Analysis team will collaborate with the 4SDI to evaluate the various new emission models and their capabilities. The identification and implementation of improved emissions modeling is one of the team's FY 2001 goals.

12) **RECOMMENDATION:** *PNGV should conduct trade-off analyses to establish relative priorities for fuel cell technical targets and cost targets.*

RESPONSE: We concur with Recommendation #12. Trade-off analyses are being conducted through fuel cell system modeling done at Argonne National Laboratory, and through cost analyses being done by Arthur D. Little and Directed Technologies, Inc. Vehicle-level modeling is also being carried out by the National Renewable Energy Laboratory and will be coordinated with the PNGV Systems Analysis Technical Team. Specifically, trade-offs between efficiency, power density, and cost will be carefully examined before revising fuel cell targets in the PNGV Technical Roadmap.

13) **RECOMMENDATION:** *PNGV should also support efforts to apply the materials improvements achieved in the program to improve lithium battery technology and to validate improved performance, life and safety.*

RESPONSE: In FY 1999 the Electrochemical Energy Storage Technical Team established the Advanced Technology Development (ATD) program through the

Department of Energy. This program is a $7 million/year multi-national-laboratory effort focused specifically on life, abuse tolerance, and cost of lithium battery technology. Within the program one task is focused on screening the latest materials available in the commercial sector to identify those materials that possess inherent safety and high performance characteristics. Another is focused on developing advanced materials based on findings from the state-of the-art diagnostic studies. This information is being transferred to the battery developers at the regularly scheduled ATD quarterly review meetings. As a result of the program interactions, unique chemistries have been developed and are being assessed by the laboratories under the protection of non-disclosure agreements with the developers.

14) **RECOMMENDATION:** *The Electrical and Electronics Systems Technical Team should closely monitor the progress towards meeting the cost goals of the automotive integrated power module and automotive electric motor drive and update and communicate realistic expectations for costs in 2004 to the systems-analysis team.*

RESPONSE: The EE Tech Team has been seized with the problem of reducing cost of power electronics and motor drive from the day one in order to make the Hybrid Powertrain a reality. As such, we set our cost targets as the most affordable. If we compare these targets with commercial products available today they look too aggressive and the peer review panel members pointed out to us year before last. The first action we took last year was to have an in depth discussions on the basis for the cost projections with automotive experts at GM and Ford with some of the panel members.

In order to meet the targets for 2004, it is obvious that new technological challenges exist. After reviewing the new technologies proposed by the AIPM (Automotive Inverter Power Module) and AEMD (Automotive Electric Motor Drive) developers we felt more comfortable. They all promised that these cost targets could be achieved with the exception of the 42 V AIPM for parallel hybrid in the 15-20 kW range, as the currents are too high. In order to make sure that these projections are indeed achievable, which is also the concern of the NRC Peer Panel members as expressed above, we took two actions:

1. We requested all the developers to do a 'Gap Analysis' and submit a time line for incorporating the new technologies within the contract framework.
2. We also requested DOE to fund two projects one related to the study of the 'cost drivers of electric motor drives and power electronics' and another 'appropriate materials and manufacturing technologies needed to meet the cost targets of filter capacitors'.

Once we have the resulted from these actions available to us we will be able to update and communicate realistic expectations of costs in 2004 to the system-analysis team as well as the NRC Peer Panel members. We expect to get some preliminary results of 'Gap Analysis' by the end of this year. The two study projects have not been funded this year due to budget constraints. We might be able to kick off these studies with 2001 budget as and when it is approved. We are working closely with National Labs on this subject.

15) **RECOMMENDATION:** *The Electrical and Electronics Systems Technical Team, in collaboration with the Vehicle Engineering Team, should undertake a comprehensive study to identify the electrical load requirements of the accessory system. Although the details of the accessory system will differ among the three USCAR partners, the impact of the accessory load is important enough that it should be considered explicitly by the system analysis team.*

RESPONSE: The vehicle electrical load is an important consideration in the overall synthesis of the PNGV vehicle. This was recognized early on by both the EE team as well as the Systems Analysis team. Much effort was placed on this area during the first year of activity in order to address the overall electrical accessory system for the vehicle. It became apparent that the accessory load is very highly depended not only on the specific design intent for the vehicle, but how these loads are managed throughout the vehicle. Each of these can have a substantial influence on the magnitude of these accessory loads. In order to address this issue within the collaborative body of PNGV a "typical" system was configured and subsequently included in the EE and Systems teams analysis. The specifications utilized for the Electrical Team contract efforts include this definition. And the System analysis tools, both PSAT and Advisor are capable of detailed simulation and analysis of the accessory system and have been used within the OEM proprietary programs to aid in vehicle synthesis

Also the magnitude of the accessory loads influence on the vehicle fuel efficiency was recognized by the Vehicle Engineering team and the resultant work on alternative approaches to minimizing interior thermal losses (loads) was reviewed with the panel. We believe that focus has been placed on this area as evidenced by both the information presented at the NRC reviews as well as the resultant embodiments displayed within the three concept vehicles presents this spring. For further detail we would refer you to the review of the concept vehicles in the areas of:

- Electric accessory drives
- Thermally efficient interiors
- High efficient HVAC Systems
- Off cycle fuel economy

16) **RECOMMENDATION:** *At this stage of the program, PNGV should direct its program toward an appropriate compromise between fuel economy and cost using the best available technology to ensure that a market-acceptable production-prototype vehicle can be achieved by 2004 that meets Tier 2 emission standards.*

RESPONSE: Cost issues will be discussed during the Peer Review, Dec. 7-8, 2000.

17) **RECOMMENDATION:** *Defining automotive system/fuels trade-offs and establishing the basis for planning for supplying required fuels as higher efficiency vehicles become commercially available will require extensive cooperation among automotive and petroleum industry representatives at all levels of responsibility. PNGV should expand and strengthen its cooperative efforts with the petroleum industry, including issues related to fuels for fuel cells. Government leadership will*

be necessary to initiate this cooperative effort and provide incentives for petroleum company involvement.

RESPONSE: We are having discussions about having a joint symposium with the petroleum industry on fuels. Government cooperation and participation will be critical.

18) **RECOMMENDATION:** *PNGV should undertake a study to assess the opportunities and costs for generating hydrogen for fuel cells at existing service stations and storing it on board vehicles and compare the feasibility, efficiency, and safety of this option with onboard fuel reforming. This study will help PNGV determine how much additional effort should be devoted to the development of onboard fuel reforming technologies.*

RESPONSE: As part of the fuels task group efforts, we are considering hydrogen as a fuel and examining the safety aspects of such a choice.

Appendix C

Presentations and Committee Activities

1. Committee Meeting, December 7–8, 2000, USCAR Headquarters, Southfield, Michigan.

Introductions
Craig Marks, Committee Chair
Larry Burns, Vice President, General Motors Research, Development & Planning

Discussion of Peer Review 6—Comments and Recommendations
PNGV Executive Committee

Systems Analysis Team
Mutasim Salman, General Motors
David Wernetts, Ford
Richard Swiatek, DaimlerChrysler

Manufacturing/Materials/Safety Working Group/Vehicle Engineering Tech Teams
Bill Charron, Ford
Manish Mehta, National Center for Manufacturing Sciences
Jim Quinn, General Motors
David Wagner, Ford
Bill Shepard, General Motors

4SDI Tech Team/HCCI/Emission Control/Fuels
Ken Howden, U.S. Department of Energy
Tom Asmus, DaimlerChrysler
Bob Carling, Sandia National Laboratories
Richard Belaire, Ford
Loren Beard, DaimlerChrysler

Fuel Cell Technical Team
JoAnn Milliken, U.S. Department of Energy
Romesh Kumar, Argonne National Laboratory

Electrical/Electronics Technical Team
Balarama Murty, General Motors
Alex Gibson, Ford
Christopher Willner, DaimlerChrysler

Battery Technical Team
Tom Tartamella, DaimlerChrysler

Summary of Highlights and Major Issues
Al Murray, PNGV Technical Manager, Ford
Jerry Rogers, PNGV Technical Manager, General Motors
Owen Viergutz, PNGV Technical Manager, DaimlerChrysler

2. Committee Meeting, February 22–24, 2001, USCAR Headquarters, Southfield, Michigan.

Opening Remarks
Craig Marks, Committee Chair

Update on California Fuel Cell Partnership
John Wallace, Th!nk Group and Ford Motor Company

Goal 1 Successes—Progress and Anticipated Efforts
Bill Charron, Ford

Government NIST and ATP Successes
Richard Ricker, National Institute of Standards and Technology

Goal 2 Report Overview
Bob Culver, USCAR Executive Director

Ford Proprietary Information-Gathering Session
Mike Schwarz, PNGV Director
Rich Belair, Bill Charron, Christine Lambert, Andy Sherman, Mike Tamor, David Wagner, Tom Watson

DOE PNGV R&D Plan for 2004
Steve Chalk, Office of Advanced Automotive Technologies, U.S. Department of Energy

PNGV Assessment of Japanese Hybrid Production Vehicles
Dan Santini and Mike Duoba, Argonne National Laboratory

Technical Cost Targets for Individual Components, Total Vehicle Systems, and Update on PNGV Vehicle Cost Model Development Process
Jerry Rogers, General Motors

Process for Monitoring Unsolicited Ideas
Bob Culver, USCAR Executive Director
Ed Wall, PNGV Coordinator, U.S. Department of Energy

DaimlerChrysler Proprietary Information-Gathering Session
Steve Zimmer, PNGV Director
Owen Viergutz, PNGV Technical Manager

Developments in High-Speed Direct Injection Engines for Light-Duty Applications in Europe
Dean Tomazic, FEV Engine Technology, Inc.

General Motors Proprietary Information-Gathering Session
Ron York, PNGV Director
Jerry Rogers, PNGV Technical Manager
David Grossman, Marty Freeman

3. Committee Meeting, April 19–20, 2001, USCAR Headquarters, Southfield, Michigan.

Review of the 2002 Budget for PNGV-Related Activities
Cary Gravatt, U.S. Department of Commerce

Washington Perspective on PNGV
Mark Kemer, General Motors

Appendix D

United States Council for Automotive Research Consortia

The U.S. automotive industry, through USCAR, has implemented collaborative projects that directly or indirectly support PNGV objectives. The USCAR consortia are listed below.

- Automotive Composites Consortium
- Electrical Wiring Component Applications Partnership
- Environmental Research Consortium
- Low Emissions Paint Consortium
- Low Emissions Technologies R&D Partnership
- Natural Gas Vehicle Technology Partnership
- Occupant Safety Research Partnership
- Supercomputer Automotive Applications Partnership
- United States Advanced Battery Consortium
- United States Automotive Materials Partnership
- Vehicle Recycling Partnership

Acronyms and Abbreviations

ac	alternating current
ACC	Advanced Composites Consortium
AEMD	automotive electric motor drive
AIPM	automotive integrated power module
AISI	American Iron and Steel Institute
Al MCC	aluminum metal matrix composite
ANL	Argonne National Laboratory
APBF-DEC	Advanced Petroleum-Based Fuel–Diesel Emission Control
ASM	American Society of Metals
AVFL	Advanced Vehicle, Fuel, Lubricant [Committee]
BIW	body-in-white
BNL	Brookhaven National Laboratory
BTU	British thermal unit
CAD	computer-aided design
CAFE	corporate average fuel economy
CARB	California Air Resources Board
CFRP	carbon-fiber-reinforced polymer
CIDI	compression-ignition direct-injection
CO	carbon monoxide
CO_2	carbon dioxide
CRADA	cooperative research and development agreement
CRC	Coordinating Research Council
CRT	continually regenerating trap

dc	direct current
DECSE	Diesel Emission Control Sulfur Effects
DIATA	direct-injection, aluminum, through-bolt assembly
DME	dimethyl ether
DMM	dimethoxymethane
DOE	U.S. Department of Energy
EE Tech Team	Electrical and Electronics Systems Technical Team
EES	electrochemical energy storage
EGR	exhaust-gas recirculation
EPA	Environmental Protection Agency
EPRI	Electric Power Research Institute
ESD	Engineering Society of Detroit
EUCAR	European Council for Automotive Research and Development
FEA	finite element analysis
FMVSS	Federal Motor Vehicle Safety Standard
FRP	fiber-reinforced polymer
FT	Fischer-Tropsch
FTP	Federal Test Procedure
GFRP	glass-fiber-reinforced polymer
GM	General Motors Corporation
HC	hydrocarbon
HCCI	homogeneous charge compression ignition
HEV	hybrid-electric vehicle
HHV	higher heating value
ICE	internal combustion engine
IFC	International Fuel Cells
IGBT	insulated gate bipolar transistor
IIHS	Insurance Institute for Highway Safety
IMA	integrated motor assist
kWe	kilowatt electric
LANL	Los Alamos National Laboratory
LEV	low-emission vehicle
LHV	lower heating value
Li-ion	lithium ion
LIMBT	lightweight injection-molded glass-fiber-reinforced plastic body technology
LIP	lithium ion polymer

LSHC	low-sulfur low-aromatics high-cetane petroleum-based diesel
MMC	metal matrix composite
mpg	miles per gallon
MTBE	methyl tertiary-butyl ether
NASA	National Aeronautics and Space Administration
NHTSA	National Highway Traffic Safety Administration
NiMH	nickel metal hydride
NO_x	nitrogen oxides
NREL	National Renewable Energy Laboratory
NSF	National Science Foundation
OAAT	Office of Advanced Automotive Technologies
ORNL	Oak Ridge National Laboratory
PAH	polyaromatic hydrocarbon
PEM	proton exchange membrane
PM	particulate matter
PNGV	Partnership for a New Generation of Vehicles
PNNL	Pacific Northwest National Laboratory
POX	partial oxidation
ppm	parts per million
PSAT	PNGV Systems Analysis Tool
R&D	research and development
SAE	Society of Automotive Engineers
SBIR	small business innovation research program
SCR	selective catalytic reduction
SNL	Sandia National Laboratories
SO_2	sulfur dioxide
SPCO	Silicon Power Corporation
SUV	sport utility vehicle
TACOM	U.S. Army Tank Automotive Command
ULEV	ultralow emission vehicle
ULSAB	ultralight steel auto body
ULSAB-AVC	Ultralight Steel Auto Body-Advanced Vehicle Concepts
USABC	United States Advanced Battery Consortium
USCAR	United States Council for Automotive Research
4SDI	four-stroke direct-injection